AF556156

Environmental Science
A Global Concern

Environmental Science
A Global Concern

Guru Sharan Das

RANDOM PUBLICATIONS
NEW DELHI (INDIA)

Environmental Science A Global Concern

ISBN 978-93-5111-527-4
© Reserved

All Rights Reserved. No Part of this book may be reproduced in any manner without written permission.
Published in 2015 in India by

RANDOM PUBLICATIONS

4376-A/4B, Gali Murari Lal, Ansari Road
New Delhi-110 002
Phone : +9111-43580356, 011-23289044, 011-43142548
e-mail: sales@randompublications.com,
info@randompublications.com, randomexports@gmail.com

Reprinted 2023

Type Setting by : Friends Media, Delhi-110089
Digitally Printed at : Replika Press Pvt. Ltd.

Preface

Environment influence and shaped our life. It is from the environment that we get food to eat, water to drink, air to breathe and all necessities of day to day life are available from our environment. Thus it is the life support system. Hence the scope and importance of the environment can be well understood. The basic concepts of environmental science are interesting and important too not only to the scientist's engaged in various fields of science and technology but also to the personnel involved in resource planning and material management. It is now universally realised that any future developmental activities have to be viewed in the light of its ultimate environmental impact.

The tremendous increase in industrial activity during the last few decades and the release of obnoxious industrial wastes in to the environment, have been considerable concern in recent years from the point of view of the environmental pollution. Environmental pollution on one hand and deforestation, soil erosion, population explosion, global warming inference in ecosystem and biosphere on the other are threatening the very existence of life on the earth.

Since human life is dependable upon the sustainable environment, its absence causes many adverse and harmful effects. Many countries are facing the problems of environmental pollutions. Without sustainable environment and natural resources, one cannot survive. Therefore, environmental aspect should be in mind while planning for industries, township and other research institutions/health centres.

The present book provides readers with an up-to-date, introductory view of essential themes in environmental science. It is intended for use in courses in environmental science, human ecology, or environmental studies the college level.

Author

Contents

1

Basics of Environmental Science

Environmental science is an interdisciplinary academic field that integrates physical and biological sciences to the study of the environment, and the solution of environmental problems. Environmental science provides an integrated, quantitative, and interdisciplinary approach to the study of environmental systems.

Related areas of study include environmental studies and environmental engineering. Environmental studies incorporates more of the social sciences for understanding human relationships, perceptions and policies towards the environment. Environmental engineering focuses on design and technology for improving environmental quality.

Environmental scientists work on subjects like the understanding of earth processes, evaluating alternative energy systems, pollution control and mitigation, natural resource management, and the effects of global climate change. Environmental issues almost always include an interaction of physical, chemical, and biological processes. Environmental scientists bring a systems approach to the analysis of environmental problems. Key elements of an effective environmental scientist include the ability to relate space, and time relationships as well as quantitative analysis.

Environmental science came alive as a substantive, active field of scientific investigation in the 1960s and 1970s driven by (a) the need for a multi-disciplinary approach to analyze complex environmental problems, (b) the arrival of substantive environmental laws requiring specific environmental protocols of investigation and (c) the growing public awareness of a need for action in addressing environmental problems. Events that spurred this development included the publication of Rachael Carson's landmark environmental book Silent Spring along with major environmental issues becoming very public, such as the 1969 Santa Barbara oil spill, and the Cuyahoga River of Cleveland, Ohio, "catching fire" (also in 1969), and helped increase the visibility of environmental issues and create this new field of study.

Atmospheric Sciences

Atmospheric sciences is an umbrella term for the study of the atmosphere, its processes, the effects other systems have on the atmosphere, and the effects of the atmosphere on these other systems. Meteorology includes atmospheric chemistry and atmospheric physics with a major focus on weather forecasting. Climatology is the study of atmospheric changes (both long and short-term) that define average climates and their change over time, due to both natural and anthropogenic climate variability. Aeronomy is the study of the upper layers of the atmosphere, where dissociation and ionization are important. Atmospheric science has been extended to the field of planetary science and the study of the atmospheres of the planets of the solar system.

Atmospheric Chemistry

Atmospheric chemistry is a branch of atmospheric science in which the chemistry of the Earth's atmosphere and that of other planets is studied. It is a multidisciplinary field of research and draws on environmental chemistry, physics, meteorology, computer modeling, oceanography, geology and volcanology and other disciplines. Research is increasingly connected with other areas of study such as climatology.

The composition and chemistry of the atmosphere is of importance for several reasons, but primarily because of the interactions between the atmosphere and living organisms. The composition of the Earth's atmosphere has been changed by human activity and some of these changes are harmful to human health, crops and ecosystems. Examples of problems which have been addressed by atmospheric chemistry include acid rain, photochemical smog and global warming. Atmospheric chemistry seeks to understand the causes of these problems, and by obtaining a theoretical understanding of them, allow possible solutions to be tested and the effects of changes in government policy evaluated.

Atmospheric Dynamics

Atmospheric dynamics involves the study of observations and theory dealing with all motion systems of meteorological importance. The list includes diverse phenomena as thunderstorms, tornadoes, gravity waves, tropical cyclones, extratropical cyclones, jet streams, and global-scale circulations. The goal of dynamical studies is to explain the observed circulations on the basis of fundamental principles from physics. The objectives of such studies include improving weather forecasting, developing methods for predicting seasonal and interannual climate fluctuations, and understanding the implications of human-induced perturbations (e.g., increased carbon dioxide concentrations or depletion of the ozone layer) on the global climate.

Atmospheric Physics

Atmospheric physics is the application of physics to the study of the atmosphere. Atmospheric physicists attempt to model Earth's atmosphere and the atmospheres of the other planets using fluid flow equations, chemical models, radiation balancing, and energy transfer processes in the atmosphere and underlying oceans. In order to model weather systems, atmospheric physicists employ elements of scattering theory, wave propagation models, cloud physics, statistical mechanics and spatial statistics which are highly mathematical and related to physics. It has close links to meteorology and climatology and also covers the design and construction of instruments for studying the atmosphere and the interpretation of the data they provide, including remote sensing instruments.

In the UK, atmospheric studies are underpinned by the Meteorological Office. Divisions of the U.S. National Oceanic and Atmospheric Administration (NOAA) oversee research projects and weather modeling involving atmospheric physics. The U.S. National Astronomy and Ionosphere Center also carries out studies of the high atmosphere. The Earth's magnetic field and the solar wind interact with the atmosphere, creating the ionosphere, Van Allen radiation belts, telluric currents, and radiant energy.

Climatology

In contrast to meteorology, which studies short term weather systems lasting up to a few weeks, climatology studies the frequency and trends of those systems. It studies the periodicity of weather events over years to millennia, as well as changes in long-term average weather patterns, in relation to atmospheric conditions. Climatologists, those who practice climatology, study both the nature of climates – local, regional or global – and the natural or human-induced factors that cause climates to change. Climatology considers the past and can help predict future climate change.

Phenomena of climatological interest include the atmospheric boundary layer, circulation patterns, heat transfer (radiative, convective and latent), interactions between the atmosphere and the oceans and land surface (particularly vegetation, land use and topography), and the chemical and physical composition of the atmosphere. Related disciplines include astrophysics, atmospheric physics, chemistry, ecology, physical geography, geology, geophysics, glaciology, hydrology, oceanography, and volcanology.

Atmospheres on Other Planets

All of the Solar System planets have atmospheres as their large masses mean gravity is strong enough to keep gaseous particles close to the surface. The larger gas giants are massive enough to keep large amounts of the light gases hydrogen and helium close by, while the smaller planets lose these gases into space. The composition of the Earth's atmosphere is different from the other planets because the various life processes that have transpired on the planet have introduced free molecular oxygen. The only solar planet without a true atmosphere is Mercury which had it mostly, although not entirely, blasted away by the solar wind. The only moon that has retained a dense atmosphere is Titan. There is a thin atmosphere on Triton, and a trace of an atmosphere on the Moon.

Planetary atmospheres are affected by the varying degrees of energy received from either the Sun or their interiors, leading to the formation of dynamic weather systems such as hurricanes, (on Earth), planet-wide dust storms (on Mars), an Earth-sized anticyclone on Jupiter (called the Great Red Spot), and holes in the atmosphere (on Neptune). At least one extrasolar planet, HD 189733 b, has been claimed to possess such a weather system, similar to the Great Red Spot but twice as large.

Hot Jupiters have been shown to be losing their atmospheres into space due to stellar radiation, much like the tails of comets. These planets may have vast differences in temperature between their day and night sides which produce supersonic winds, although the day and night sides of HD 189733b appear to have very similar temperatures, indicating that planet's atmosphere effectively redistributes the star's energy around the planet.

Ecology

Ecology is the scientific study of the relationships that living organisms have with each other and with their natural environment. Topics of interest to ecologists include the composition, distribution, amount (biomass), number, and changing states of organisms within and among ecosystems. Ecosystems are composed of dynamically interacting parts including organisms, the communities they make up, and the non-living components of their environment. Ecosystem processes, such as primary production, pedogenesis, nutrient cycling, and various niche construction activities, regulate the flux of energy and matter through an environment. These processes are sustained by the biodiversity within them. Biodiversity refers to the varieties of species in ecosystems, the genetic variations they contain, and the processes that are functionally enriched by the diversity of ecological interactions.

Ecology is an interdisciplinary field that includes biology and Earth science. The word "ecology" was coined in 1866 by the German scientist Ernst Haeckel (1834–1919). Ancient Greek philosophers such as Hippocrates and Aristotle laid the foundations of ecology in their studies on natural history. Modern ecology transformed into a more rigorous science in the late 19th century. Evolutionary concepts on adaptation and natural selection became cornerstones of modern ecological theory. Ecology is not synonymous with environment, environmentalism, natural history, or environmental science. It is closely related to evolutionary biology, genetics, and ethology. An understanding of how biodiversity affects ecological function is an important focus area in ecological studies

The scope of ecology covers a wide array of interacting levels of organization spanning micro-level (e.g., cells) to planetary scale (e.g., ecosphere) phenomena. Ecosystems, for example, contain populations of individuals that aggregate into distinct ecological communities. It can take thousands of years for ecological processes to bring about the final successional stages of a forest. An ecosystem's area can vary greatly, from tiny to vast. A single tree is of little consequence to the classification of a forest ecosystem, but critically relevant to organisms living in and on it. Several generations of an aphid population can exist over the lifespan of a single leaf. Each of those aphids, in turn, support diverse bacterial communities. The nature of connections in ecological communities cannot be explained by knowing the details of each species in isolation, because the emergent pattern is neither revealed nor predicted until the ecosystem is studied as an integrated whole. Some ecological principles, however, do exhibit collective properties where the sum of the components explain the properties of the whole, such as birth rates of a population being equal to the sum of individual births over a designated time frame.

Hierarchical Ecology

The scale of ecological dynamics can operate like a closed system, such as aphids migrating on a single tree, while at the same time remain open with regard to broader scale influences, such as atmosphere or climate. Hence, ecologists classify ecosystems hierarchically by analyzing data collected from finer scale units, such as vegetation associations, climate, and soil types, and integrate this information to identify emergent patterns of uniform organization and processes that operate on local to regional, landscape, and chronological scales.

Biodiversity

Biodiversity describes the diversity of life from genes to ecosystems and spans every level of biological organization. The term has several interpretations, and there are many ways to index, measure, characterize, and represent its complex organization. Biodiversity includes species diversity, ecosystem diversity, genetic diversity and the complex processes operating at and among these respective levels. Biodiversity plays an important role in ecological health as much as it does for human health. Preventing species extinctions is one way to preserve biodiversity, but factors such as genetic diversity and migration routes are equally important and are threatened on global scales. Conservation priorities and management techniques require different approaches and considerations to address the full ecological scope of biodiversity. Populations and species migration, for example, are sensitive indicators of ecosystem services that sustain and contribute natural capital toward the well-being of humanity. An understanding of biodiversity has practical application for ecosystem-based conservation planners as they make ecologically responsible decisions in management recommendations to consultant firms, governments, and industry. The protected areas have been established under the protected area network across the world for conservation of biodiversity.

Habitat

The habitat of a species describes the environment over which a species is known to occur and the type of community that is formed as a result. More specifically, "habitats can be defined as regions in environmental space that are composed of multiple dimensions, each representing a biotic or abiotic environmental variable; that is, any component or characteristic of the environment related directly (e.g. forage biomass and quality) or indirectly (e.g. elevation) to the use of a location by the animal.":745 For example, a habitat might be an aquatic or terrestrial environment that can be further categorized as a montane or alpine ecosystem. Habitat shifts provide important evidence of competition in nature where one population changes relative to the habitats that most other individuals of the species occupy. For example, one population of a species of tropical lizards (Tropidurus hispidus) has a flattened body relative to the main populations that live in open savanna. The population that lives in an isolated rock outcrop hides in crevasses where its flattened body offers a selective advantage. Habitat shifts also occur in the developmental life history of amphibians and in insects that transition from aquatic to terrestrial habitats. Biotope and habitat are sometimes used interchangeably, but the former applies to a community's environment, whereas the latter applies to a species' environment.

Niche

Definitions of the niche date back to 1917, but G. Evelyn Hutchinson made conceptual advances in 1957 by introducing a widely adopted definition: "the set of biotic and abiotic conditions in which a species is able to persist and maintain stable population sizes.":519 The ecological niche is a central concept in the ecology of organisms and is sub-divided into the fundamental and the realized niche. The fundamental niche is the set of environmental conditions under which a species is able to persist. The realized niche is the set of environmental plus ecological conditions under which a species persists.

Biogeographical patterns and range distributions are explained or predicted through knowledge of a species' traits and niche requirements. Species have functional traits that are uniquely adapted to the ecological niche. A trait is a measurable property, phenotype, or characteristic of an organism that may influence its survival. Genes play an important role in the interplay of development and environmental expression of traits. Resident species evolve traits that are fitted to the selection pressures of their local environment. This tends to afford them a competitive advantage and discourages similarly adapted species from having an overlapping geographic range. The competitive exclusion principle states that two species cannot coexist indefinitely by living off the same limiting resource; one will always outcompete the other. When similarly adapted species overlap geographically, closer inspection reveals subtle ecological differences in their habitat or dietary requirements. Some models and empirical studies, however, suggest that disturbances can stabilize the coevolution and shared niche occupancy of similar species inhabiting species-rich communities. The habitat plus the niche is called the ecotope, which is defined as the full range of environmental and biological variables affecting an entire species.

Organisms are subject to environmental pressures, but they also modify their habitats. The regulatory feedback between organisms and their environment can affect conditions from local (e.g., a beaver pond) to global scales, over time and even after death, such as decaying logs or silica skeleton deposits from marine organisms. The process and concept of ecosystem engineering has also been called niche construction. Ecosystem engineers are defined as: "organisms that directly or indirectly modulate the availability of resources to other species, by causing physical state changes in biotic or abiotic materials. In so doing they modify, maintain and create habitats."

The ecosystem engineering concept has stimulated a new appreciation for the influence that organisms have on the ecosystem and evolutionary process. The term "niche construction" is more often used in reference to the under-appreciated feedback mechanism of natural selection imparting forces on the abiotic niche. An example of natural selection through ecosystem engineering occurs in the nests of social insects, including ants, bees, wasps, and termites. There is an emergent homeostasis or homeorhesis in the structure of the nest that regulates, maintains and defends the physiology of the entire colony. Termite mounds, for example, maintain a constant internal temperature through the design of air-conditioning chimneys. The structure of the nests themselves are subject to the forces of natural selection. Moreover, a nest can survive over successive generations, so that progeny inherit both genetic material and a legacy niche that was constructed before their time.

Biome

Biomes are larger units of organization that categorize regions of the Earth's ecosystems, mainly according to the structure and composition of vegetation. There are different methods to define the continental boundaries of biomes dominated by different functional types of vegetative communities that are limited in distribution by climate, precipitation, weather and other environmental variables. Biomes include tropical rainforest, temperate broadleaf and mixed forest, temperate deciduous forest, taiga, tundra, hot desert, and polar desert. Other researchers have recently categorized other biomes, such as the human and oceanic microbiomes. To a microbe, the human body is a habitat and a landscape. Microbiomes were discovered largely through advances in molecular genetics, which have revealed a hidden richness of microbial diversity on the planet. The oceanic microbiome plays a significant role in the ecological biogeochemistry of the planet's oceans.

Biosphere

The largest scale of ecological organization is the biosphere: the total sum of ecosystems on the planet. Ecological relationships regulate thc flux of energy, nutrients, and climate all the way up to the planetary scale. For example, the dynamic history of the planetary atmosphere's CO2 and O2 composition has been affected by the biogenic flux of gases coming from respiration and photosynthesis, with levels fluctuating over time in relation to the ecology and evolution of plants and animals. Ecological theory has also been used to explain self-emergent regulatory phenomena at the planetary scale: for example, the Gaia hypothesis is an example of holism applied in ecological theory. The Gaia hypothesis states that there is an emergent feedback loop generated by the metabolism of living organisms that maintains the temperature of the Earth and atmospheric conditions within a narrow self-regulating range of tolerance.

Population Ecology

Population ecology studies the dynamics of species populations and how these populations interact with the environment. A population consists of individuals of the same species that live, interact and migrate through the same niche and habitat.A primary law of population ecology is the Malthusian growth model which states, "a population will grow (or decline) exponentially as long as the environment experienced by all individuals in the population remains constant.":18 Simplified population models usually start with four variables: death, birth, immigration, and emigration.

Metapopulations and migration

The concept of metapopulations was defined in 1969 as "a population of populations which go extinct locally and recolonize.":105 Metapopulation ecology is another statistical approach that is often used in conservation research. Metapopulation models simplify the landscape into patches of varying levels of quality, and metapopulations are linked by the migratory behaviours of organisms. Animal migration is set apart from other kinds of movement because it involves the seasonal departure and return of individuals from a habitat. Migration is also a population-level phenomenon, as with the migration routes followed by plants as they occupied northern post-glacial environments. Plant ecologists use pollen records that accumulate and stratify in wetlands to reconstruct the timing of plant migration and dispersal relative to historic and contemporary climates. These migration routes involved an expansion of the range as plant populations expanded from one area to another. There is a larger taxonomy of movement, such as commuting, foraging, territorial behaviour, stasis, and ranging. Dispersal is usually distinguished from migration because it involves the one way permanent movement of individuals from their birth population into another population.

In metapopulation terminology, migrating individuals are classed as emigrants (when they leave a region) or immigrants (when they enter a region), and sites are classed either as sources or sinks. A site is a generic term that refers to places where ecologists sample populations, such as ponds or defined sampling areas in a forest. Source patches are productive sites that generate a seasonal supply of juveniles that migrate to other patch locations. Sink patches are unproductive sites that only receive migrants; the population at the site will disappear unless rescued by an adjacent source patch or environmental conditions become more favourable. Metapopulation models examine patch dynamics over time to answer questions about spatial and demographic ecology. The ecology of metapopulations is a dynamic process of extinction and colonization. Small patches of lower quality (i.e., sinks) are maintained or rescued by a seasonal influx of new immigrants. A dynamic metapopulation structure evolves from year to year, where some patches are sinks in dry years and are sources when conditions are more favourable. Ecologists use a mixture of computer models and field studies to explain metapopulation structure.

Community Ecology

Community ecology is the study of the interactions among a collections of species that inhabit the same geographic area. Research in community ecology might measure primary production in a wetland in relation to decomposition and consumption rates. This requires an understanding of the community connections between plants (i.e., primary producers) and the decomposers (e.g., fungi and bacteria), or the analysis of predator-prey dynamics affecting amphibian biomass. Food webs and trophic levels are two widely employed conceptual models used to explain the linkages among species.

Ecosystem Ecology

Ecosystems are habitats within biomes that form an integrated whole and a dynamically responsive system having both physical and biological complexes. The underlying concept can be traced back to 1864 in the published work of George Perkins Marsh ("Man and Nature"). Within an ecosystem, organisms are linked to the physical and biological components of their environment to which they are adapted. Ecosystems are complex adaptive systems where the interaction of life processes form self-organizing patterns across different scales of time and space. Ecosystems are broadly categorized as terrestrial, freshwater, atmospheric, or marine. Differences stem from the nature of the unique physical environments that shapes the biodiversity within each. A more recent addition to ecosystem ecology are technoecosystems, which are affected by or primarily the result of human activity.

Food webs

A food web is the archetypal ecological network. Plants capture solar energy and use it to synthesize simple sugars during photosynthesis. As plants grow, they accumulate nutrients and are eaten by grazing herbivores, and the energy is transferred through a chain of organisms by consumption. The simplified linear feeding pathways that move from a basal trophic species to a top consumer is called the food chain. The larger interlocking pattern of food chains in an ecological community creates a complex food web. Food webs are a type of concept map or a heuristic device that is used to illustrate and study pathways of energy and material flows.

Food webs are often limited relative to the real world. Complete empirical measurements are generally restricted to a specific habitat, such as a cave or a pond, and principles gleaned from food web microcosm studies are extrapolated to larger systems. Feeding relations require extensive investigations into the gut contents of organisms, which can be difficult to decipher, or stable isotopes can be used to trace the flow of nutrient diets and energy through a food web. Despite these limitations, food webs remain a valuable tool in understanding community ecosystems.

Food webs exhibit principles of ecological emergence through the nature of trophic relationships: some species have many weak feeding links (e.g., omnivores) while some are more specialized with fewer stronger feeding links (e.g., primary predators). Theoretical and empirical studies identify non-random emergent patterns of few strong and many weak linkages that explain how ecological communities remain stable over time. Food webs are composed of subgroups where members in a community are linked by strong interactions, and the weak interactions occur between these subgroups. This increases food web stability. Step by step lines or relations are drawn until a web of life is illustrated.

Trophic levels

A trophic level is "a group of organisms acquiring a considerable majority of its energy from the adjacent level nearer the abiotic source.":383 Links in food webs primarily connect feeding relations or trophism among species. Biodiversity within ecosystems can be organized into trophic pyramids, in which the vertical dimension represents feeding relations that become further removed from the base of the food chain up toward top predators, and the horizontal dimension represents the abundance or biomass at each level. When the relative abundance or biomass of each species is sorted into its respective trophic level, they naturally sort into a 'pyramid of numbers'.

Species are broadly categorized as autotrophs (or primary producers), heterotrophs (or consumers), and Detritivores (or decomposers). Autotrophs are organisms that produce their own food (production is greater than respiration) by photosynthesis or chemosynthesis. Heterotrophs are organisms that must feed on others for nourishment and energy (respiration exceeds production). Heterotrophs can be further sub-divided into different functional groups, including primary consumers (strict herbivores), secondary consumers (carnivorous predators that feed exclusively on herbivores) and tertiary consumers (predators that feed on a mix of herbivores and predators). Omnivores do not fit neatly into a functional category because they eat both plant and animal tissues. It has been suggested that omnivores have a greater functional influence as predators, because compared to herbivores they are relatively inefficient at grazing.

Trophic levels are part of the holistic or complex systems view of ecosystems. Each trophic level contains unrelated species that are grouped together because they share common ecological functions, giving a macroscopic view of the system. While the notion of trophic levels provides insight into energy flow and top-down control within food webs, it is troubled by the prevalence of omnivory in real ecosystems. This has led some ecologists to "reiterate that the notion that species clearly aggregate into discrete, homogeneous trophic levels is fiction.":815 Nonetheless, recent studies have shown that real trophic levels do exist, but "above the herbivore trophic level, food webs are better characterized as a tangled web of omnivores."

Keystone species

A keystone species is a species that is connected to a disproportionately large number of other species in the food-web. Keystone species have lower levels of biomass in the trophic pyramid relative to the importance of their role. The many connections that a keystone species holds means that it maintains the organization and structure of entire communities. The loss of a keystone species results in a range of dramatic cascading effects that alters trophic dynamics, other food web connections, and can cause the extinction of other species.

Sea otters (Enhydra lutris) are commonly cited as an example of a keystone species because they limit the density of sea urchins that feed on kelp. If sea otters are removed from the system, the urchins graze until the kelp beds disappear and this has a dramatic effect on community structure. Hunting of sea otters, for example, is thought to have indirectly led to the extinction of the Steller's Sea Cow (Hydrodamalis gigas). While the keystone species concept has been used extensively as a conservation tool, it has been criticized for being poorly defined from an operational stance. It is difficult to experimentally determine what species may hold a keystone role in each ecosystem. Furthermore, food web theory suggests that keystone species may not be common, so it is unclear how generally the keystone species model can be applied.

Ecological Complexity

Complexity is understood as a large computational effort needed to piece together numerous interacting parts exceeding the iterative memory capacity of the human mind. Global patterns of biological diversity are complex. This biocomplexity stems from the interplay among ecological processes that operate and influence patterns at different scales that grade into each other, such as transitional areas or ecotones spanning landscapes. Complexity stems from the interplay among levels of biological organization as energy and matter is integrated into larger units that superimpose onto the smaller parts. "What were wholes on one level become parts on a higher one.":209 Small scale patterns do not necessarily explain large scale phenomena, otherwise captured in the expression (coined by Aristotle) 'the sum is greater than the parts'.

"Complexity in ecology is of at least six distinct types: spatial, temporal, structural, process, behavioral, and geometric.":3 From these principles, ecologists have identified emergent and self-organizing phenomena that operate at different environmental scales of influence, ranging from molecular to planetary, and these require different explanations at each integrative level. Ecological complexity relates to the dynamic resilience of ecosystems that transition to multiple shifting steady-states directed by random fluctuations of history. Long-term ecological studies provide important track records to better understand the complexity and resilience of ecosystems over longer temporal and broader spatial scales. These studies are managed by the International Long Term Ecological Network (LTER). The longest experiment in existence is the Park Grass Experiment, which was initiated in 1856. Another example is the Hubbard Brook study, which has been in operation since 1960.

Environmental Chemistry

Environmental chemistry is the study of chemical alterations in the environment. Principal areas of study include soil contamination and water pollution. The topics of analysis include chemical degradation in the environment, multi-phase transport of chemicals (for example, evaporation of a solvent containing lake to yield solvent as an air pollutant), and chemical effects upon biota.

As an example study, consider the case of a leaking solvent tank which has entered the habitat soil of an endangered species of amphibian. As a method to resolve or understand the extent of soil contamination and subsurface transport of solvent, a computer model would be implemented. Chemists would then characterize the molecular bonding of the solvent to the specific soil type, and biologists would study the impacts upon soil arthropods, plants, and ultimately pond-dwelling organisms that are the food of the endangered amphibian.

Geosciences

Geosciences include environmental geology, environmental soil science, volcanic phenomena and evolution of the Earth's crust. In some classification systems this can also include hydrology, including oceanography.

As an example study of soils erosion, calculations would be made of surface runoff by soil scientists. Fluvial geomorphologists would assist in examining sediment transport in overland flow. Physicists would contribute by assessing the changes in light transmission in the receiving waters. Biologists would analyze subsequent impacts to aquatic flora and fauna from increases in water turbidity.

References

Begon, M.; Townsend, C. R.; Harper, J. L. (2005). *Ecology: From Individuals to Ecosystems* (4th ed.). Wiley-Blackwell. p. 752. ISBN 1-4051-1117-8.

Odum, E. P.; Barrett, G. W. (2005). *Fundamentals of Ecology*. Brooks Cole. p. 598.

Kormondy, E. E. (1995). *Concepts of Ecology* (4th ed.). Benjamin Cummings.

McIntosh, R. P. (1985). *The Background of Ecology: Concept and Theory*. Cambridge University Press. pp. 400.

2

Environmental Chemistry

Environmental chemistry involves first understanding how the uncontaminated environment works, which chemicals in what concentrations are present naturally, and with what effects. Without this it would be impossible to accurately study the effects humans have on the environment through the release of chemicals.

Environmental chemistry is used by the Environment Agency (in England and Wales), the Environmental Protection Agency (in the United States) the Association of Public Analysts, and other environmental agencies and research bodies around the world to detect and identify the nature and source of pollutants. These can include:

— Heavy metal contamination of land by industry. These can then be transported into water flows and be taken up by living organisms.

— Nutrients such as nitrate and phosphate leaching from agricultural land into water courses, which can lead to algal blooms and eutrophication.

Quantitative chemical analysis is a key part of environmental chemistry, since it provides the data that frame most environmental studies. Common analytical techniques used for quantitative determinations in environmental chemistry include classical wet chemistry, such as gravimetric, titrimetric and electrochemical methods. More sophisticated approaches are used in the determination of trace metals and organic compounds. Metals are commonly measured by atomic spectroscopy and mass spectrometry: Atomic Absorption Spectrophotometry (AA) and Inductively Coupled Plasma

Atomic Emission (ICP-AES) or Inductively Coupled Plasma Mass Spectrometric (ICP-MS) techniques. Organic compounds are commonly measured also using mass spectrometric methods, such as Gas chromatography-mass spectrometry (GC/MS) and Liquid chromatography-mass spectrometry (LC/MS). Non-MS methods using GCs and LCs having universal or specific detectors are still staples in the arsenal of available analytical tools.

Other parameters often measured in environmental chemistry are radiochemicals. These are pollutants which emit radioactive materials, such as alpha and beta particles, posing danger to human health and the environment. Particle counters and Scintillation counters are most commonly used for these measurements. Bioassays and immunoassays are utilised for toxicity evaluations of chemical effects on various organisms.

Atmospheric Composition

Atmospheric chemistry is a branch of atmospheric science in which the chemistry of the Earth's atmosphere and that of other planets is studied. It is a multidisciplinary field of research and draws on environmental chemistry, physics, meteorology, computer modeling, oceanography, geology and volcanology and other disciplines. The composition and chemistry of the atmosphere is of importance for several reasons, but primarily because of the interactions between the atmosphere and living organisms. The composition of the Earth's atmosphere has been changed by human activity and some of these changes are harmful to human health, crops and ecosystems. Examples of problems which have been addressed by atmospheric chemistry include acid rain, photochemical smog and global warming. Atmospheric chemistry seeks to understand the causes of these problems, and by obtaining a theoretical understanding of them, allow possible solutions to be tested and the effects of changes in government policy evaluated.

The ancient Greeks regarded air as one of the four elements, but the first scientific studies of atmospheric composition began in the 18th century. Chemists such as Joseph Priestley, Antoine Lavoisier and Henry Cavendish made the first measurements of the composition of the atmosphere.

In the late 19th and early 20th centuries interest shifted towards trace constituents with very small concentrations. One particularly important discovery for atmospheric chemistry was the discovery of ozone by Christian Friedrich Schoenbein in 1840.

In the 20th century atmospheric science moved on from studying the composition of air to a consideration of how the concentrations of trace gases in the atmosphere have changed over time and the chemical processes which create and destroy compounds in the air. Two particularly important examples of this were the explanation of how the ozone layer is created and maintained by Sydney Chapman and Gordon Dobson, and the explanation of Photochemical smog by Haagen-Smit.

In the 21st century the focus is now shifting again. Atmospheric Chemistry is increasingly studied as one part of the Earth system. Instead of concentrating on atmospheric chemistry in isolation the focus is now on seeing it as one part of a single system with the rest of the atmosphere, biosphere and geosphere. An especially important driver for this is the links between chemistry and climate such as the effects of changing climate on the recovery of the ozone hole and vice versa but also interaction of the composition of the atmosphere with the oceans and terrestrial ecosystems.

Observations, lab measurements and modeling are the three central elements in atmospheric chemistry. Progress in atmospheric chemistry is often driven by the interactions between these components and they form an integrated whole. For example observations may tell us that more of a chemical compound exists than previously thought possible. This will stimulate new modelling and laboratory studies which will increase our scientific understanding to a point where the observations can be explained.

Observation

Observations of atmospheric chemistry are essential to our understanding. Routine observations of chemical composition tell us about changes in atmospheric composition over time. One important example of this is the Keeling Curve - a series of measurements from 1958 to today which show a steady rise in of the concentration of carbon dioxide. Observations of atmospheric chemistry are made in observatories such as that on Mauna Loa and on mobile platforms such as aircraft (e.g. the UK's Facility for Airborne Atmospheric Measurements), ships and balloons. Observations of atmospheric composition are increasingly made by satellites with important instruments such as GOME and MOPITT giving a global picture of air pollution and chemistry. Surface observations have the advantage that they provide long term records at high time resolution but are limited in the vertical and horizontal space they provide observations from. Some surface based instruments e.g. LIDAR can provide concentration profiles of chemical compounds and aerosol but are still restricted in the horizontal region they can cover. Many observations are available on line in Atmospheric Chemistry Observational Databases.

Lab Measurements

Measurements made in the laboratory are essential to our understanding of the sources and sinks of pollutants and naturally occurring compounds. Lab studies tell us which gases react with each other and how fast they react. Measurements of interest include reactions in the gas phase, on surfaces and in water. Also of high importance is photochemistry which quantifies how quickly molecules are split apart by sunlight and what the products are plus thermodynamic data such as Henry's law coefficients.

Modeling

In order to synthesise and test theoretical understanding of atmospheric chemistry, computer models (such as chemical transport models) are used. Numerical models solve the differential equations governing the concentrations of chemicals in the atmosphere. They can be very simple or very complicated. One common trade off in numerical models is between the number of chemical compounds and chemical reactions modelled versus the representation of transport and mixing in the atmosphere. For example, a box model might include hundreds or even thousands of chemical reactions but will only have a very crude representation of mixing in the atmosphere. In contrast, 3D models represent many of the physical processes of the atmosphere but due to constraints on computer resources will have far fewer chemical reactions and compounds. Models can be used to interpret observations, test understanding of chemical reactions and predict future concentrations of chemical compounds in the atmosphere. One important current trend is for atmospheric chemistry modules to become one part of earth system models in which the links between climate, atmospheric composition and the biosphere can be studied.

Some models are constructed by automatic code generators. In this approach a set of constituents are chosen and the automatic code generator will then select the reactions involving those constituents from a set of reaction databases. Once the reactions have been chosen the ordinary differential equations (ODE) that describe their time evolution can be automatically constructed.

Acid Rain

Acid rain is rain or any other form of precipitation that is unusually acidic. It has harmful effects on plants, aquatic animals, and infastructure. Acid rain is mostly caused by human emissions of sulfur and nitrogen compounds which react in the atmosphere to produce acids. In recent years, many governments have introduced laws to reduce these emissions.

"Acid rain" is a popular term referring to the deposition of wet (rain, snow, sleet, fog and cloudwater, dew) and dry (acidifying particles and gases) acidic components. A more accurate term is "acid deposition". Distilled water, which contains no carbon dioxide, has a neutral pH of 7. Liquids with a pH less than 7 are acidic, and those with a pH greater than 7 are basic. "Clean" or unpolluted rain has a slightly acidic pH of about 5.2, because carbon dioxide and water in the air react together to form carbonic acid, a weak acid (pH 5.6 in distilled water), but unpolluted rain also contains other chemicals.

$$H_2O\ (l) + CO_2\ (g) \rightarrow H_2CO_3\ (aq)$$

Carbonic acid then can ionise in water forming low concentrations of hydronium ions:

$$2H_2O\ (l) + H_2CO_3\ (aq) \rightarrow CO_3^{2-}\ (aq) + 2H_3O^+(aq)$$

The extra acidity in rain comes from the reaction of primary air pollutants, primarily sulfur oxides and nitrogen oxides, with water in the air to form strong acids (like sulfuric and nitric acid). The main sources of these pollutants are industrial power-generating plants and vehicles.

Since the Industrial Revolution, emissions of sulphur dioxide and nitrogen oxides to the atmosphere have increased. In 1852, Robert Angus Smith was the first to show the relationship between acid rain and atmospheric pollution in Manchester, England. Though acidic rain was discovered in 1852, it wasn't until the late 1960s that scientists began widely observing and studying the phenomenon. The term "acid rain" was generated in 1972. Canadian Harold Harvey was among the first to research a "dead" lake. Public awareness of acid rain in the U.S increased in the 1970s after the New York Times promulgated reports from the Hubbard Brook Experimental Forest in New Hampshire of the myriad deleterious environmental effects demonstrated to result from it.

Occasional pH readings in rain and fog water of well below 2.4 (the acidity of vinegar) have been reported in industrialised areas. Industrial acid rain is a substantial problem in Europe, China, Russia and areas down-wind from them. These areas all burn sulfur-containing coal to generate heat and electricity. The problem of acid rain not only has increased with population and industrial growth, but has become more widespread. The use of tall smokestacks to reduce local pollution has contributed to the spread of acid rain by releasing gases into regional atmospheric circulation. Often deposition occurs a considerable distance downwind of the emissions, with mountainous regions tending to receive the greatest deposition (simply because of their higher rainfall). An example of this effect is the low pH of rain (compared to the local emissions) which falls in Scandinavia.

Emissions of Chemicals Leading to Acidification

The most important gas which leads to acidification is sulfur dioxide. Emissions of nitrogen oxides which are oxidised to form nitric acid are of increasing importance due to stricter controls on emissions of sulfur containing compounds. 70 Tg(S) per year in the form of SO_2 comes from fossil fuel combustion and industry, 2.8 Tg(S) from wildfires and 7-8 Tg(S) per year from volcanoes.

Natural phenomena

The principal natural phenomena that contribute acid-producing gases to the atmosphere are emissions from volcanoes and those from biological processes that occur on the land, in wetlands, and in the oceans. The major biological source of sulfur containing compounds is dimethyl sulfide.

Human activity

The principal cause of acid rain is sulfur and nitrogen compounds from human sources, such as electricity generation, factories, and motor vehicles. Coal power plants are one of the most polluting. The gases can be carried hundreds of kilometres in the atmosphere before they are converted to acids and deposited. In the past, factories had short funnels to let out smoke, but this caused many problems locally; thus, factories now have taller smoke funnels. However, dispersal from these taller stacks causes pollutants to be carried farther, causing widespread ecological damage.

Acid Deposition

Wet deposition

Wet deposition of acids occurs when any form of precipitation (rain, snow, etc) removes acids from the atmosphere and delivers it to the Earth's surface. This can result from the deposition of acids produced in the raindrops or by the precipitation removing the acids either in clouds or below clouds. Wet removal of both gases and aerosols are both of importance for wet deposition.

Dry deposition

Acid deposition also occurs via dry deposition in the absence of precipitation. This can be responsible for as much as 20 to 60% of total acid deposition. This occurs when particles and gases stick to the ground, plants or other surfaces.

Adverse Effects

Acid rain has been shown to have adverse impacts on forests, freshwaters and soils, killing insect and aquatic lifeforms as well as causing damage to buildings and having impacts on human health.

Surface waters and aquatic animals

Both the lower pH and higher aluminum concentrations in surface water that occur as a result of acid rain can cause damage to fish and other aquatic animals. At pHs lower than 5 most fish eggs will not hatch and lower pHs can kill adult fish. As lakes and rivers become more acidic biodiversity is reduced. Acid rain has eliminated insect life and some fish species, including the brook trout in some lakes, streams, and creeks in geographically sensitive areas, such as the Adirondack Mountains of the United States. However, the extent to which acid rain contributes directly or indirectly via runoff from the catchment to lake and river acidity (i.e., depending on characteristics of the surrounding watershed) is variable.

Soils

Soil biology and chemistry can be seriously damaged by acid rain. Some microbes are unable to tolerate changes to low pHs and are killed. The enzymes of these microbes are denatured by the acid. The hydronium ions of acid rain also mobilise toxins, e.g. aluminium, and leach away essential nutrients and minerals.

$$2H^{+}\ (aq) + Mg^{2+}\ (clay) \rightarrow 2H^{+}\ (clay) + Mg^{2+}(aq)$$

Soil chemistry can be dramatically changed when base cations, such as calcium and magnesium, are leached by acid rain thereby affecting sensitive species, such as sugar maple (Acer saccharum).

Forests and other vegetation

Adverse effects may be indirectly related to acid rain, like the acid's effects on soil or high concentration of gaseous precursors to acid rain. High altitude forests are especially vulnerable as they are often surrounded by clouds and fog which are more acidic than rain. Other plants can also be damaged by acid rain but the effect on food crops is minimised by the application of lime and fertilisers to replace lost nutrients. In cultivated areas, limestone may also be added to increase the ability of the soil to keep the pH stable, but this tactic is largely unusable in the case of wilderness lands. When calcium is leached from the needles of red spruce, these trees become less cold tolerant and exhibit winter injury and even death.

Human health

Scientists have suggested direct links to human health. Fine particles, a large fraction of which are formed from the same gases as acid rain (sulfur dioxide and nitrogen dioxide), have been shown to cause illness and premature deaths such as cancer and other diseases.

Other adverse effects

Acid rain can also cause damage to certain building materials and historical monuments. This results when the sulfuric acid in the rain chemically reacts with the calcium compounds in the stones (limestone, sandstone, marble and granite) to create gypsum, which then flakes off.

$$CaCO_3\ (s) + H_2SO_4\ (aq) \rightarrow CaSO_4\ (aq) + CO_2\ (g) + H_2O\ (l)$$

This result is also commonly seen on old gravestones where the acid rain can cause the inscription to become completely illegible. Acid rain also causes an increased rate of oxidation for iron. Visibility is also reduced by sulfate and nitrate aerosols and particles in the atmosphere.

Affected Areas

Particularly badly affected places around the globe include most of Europe many parts of the United States and South Western Canada. Other affected areas include the South Eastern coast of China and Taiwan. Places like much of South Asia (Indonesia, Malaysia and Thailand), Western South Africa (the country), Southern India and Sri Lanka and even West Africa (countries like Ghana, Togo and Nigeria) could all be prone to acidic rainfall in the future.

Prevention Methods

Technical solutions

In the United States, many coal-burning power plants use Flue gas desulfurisation (FGD) to remove sulfur-containing gases from their stack gases. An example of FGD is the wet scrubber which is commonly used in the U.S. and many other countries. A wet scrubber is basically a reaction tower equipped with a fan that extracts hot smoke stack gases from a power plant into the tower. Lime or limestone in slurry form is also injected into the tower to mix with the stack gases and combine with the sulfur dioxide present. The calcium carbonate of the limestone produces pH-neutral calcium sulfate that is physically removed from the scrubber. That is, the scrubber turns sulfur pollution into industrial sulfates.

In some areas the sulfates are sold to chemical companies as gypsum when the purity of calcium sulfate is high. In others, they are placed in landfill. However, the effects of acid rain can last for generations, as the effects of pH level change can stimulate the continued leaching of undesirable chemicals into otherwise pristine water sources, killing off vulnerable insect and fish species and blocking efforts to restore native life. Automobile emissions control reduces emissions of nitrogen oxides from motor vehicles.

International treaties

A number of international treaties on the long range transport of atmospheric pollutants have been agreed e.g. Sulphur Emissions Reduction Protocol under the Convention on Long-Range Transboundary Air Pollution.

Emissions trading

In this regulatory scheme, every current polluting facility is given or may purchase on an open market an emissions allowance for each unit of a designated pollutant it emits. Operators can then install pollution control equipment, and sell portions of their emissions allowances they no longer need for their own operations, thereby recovering some of the capital cost of their investment in such equipment. The intention is to give operators economic incentives to install pollution controls.

The first emissions trading market was established in the United States by enactment of the Clean Air Act Amendments of 1990. The overall goal of the Acid Rain Program established by the Act is to achieve significant environmental and public health benefits through reductions in emissions of sulfur dioxide (SO_2) and nitrogen oxides (NO_x), the primary causes of acid rain. To achieve this goal at the lowest cost to society, the program employs both regulatory and market based approaches for controlling air pollution.

Greenhouse Effect

Greenhouse gases are gaseous constituents of thc atmosphcre, both natural and anthropogenic, that absorb and emit radiation at specific wavelengths within the spectrum of thermal infrared radiation emitted by the Earth's surface, the atmosphere itself, and by clouds. This property causes the greenhouse effect. Greenhouse gases are essential to maintaining the temperature of the Earth; without them the planet would be so cold as to be uninhabitable. Natural sources are the Earth's ecosystem, and anthropogenic sources include industrial, transportation, residential, commercial and agricultural processes. Venus, Mars and Titan also have atmospheric gases that cause greenhouse effects.

In order, Earth's most abundant greenhouse gases are:

— water vapor
— carbon dioxide
— methane
— nitrous oxide
— ozone
— CFCs

By effect, the most important greenhouse gases are:

- water vapor, which causes about 36–70% of the greenhouse effect on Earth.
- carbon dioxide, which causes 9–26%
- methane, which causes 4–9%
- ozone, which causes 3–7%

Importance is a combination of the strength of the greenhouse effect of the gas and its abundance. For example, methane is a much stronger greenhouse gas than carbon dioxide, but is present in much smaller concentrations. It is not possible to state that a certain gas causes a certain percentage of the greenhouse effect, because the influences of the various gases are not additive. Other greenhouse gases include, but are not limited to, nitrous oxide, sulfur hexafluoride, hydrofluorocarbons, perfluorocarbons and chlorofluorocarbons. Some greenhouse gases are present but not often listed. For example, nitrogen trifluoride has a high global warming potential (GWP) but is only present in small quantities.

The major atmospheric constituents (nitrogen, N_2 and oxygen, O_2) are not greenhouse gases. Nor is the approximately 1% of argon, Ar. This is because homonuclear diatomic molecules such as N_2 and O_2 and monatomic molecules such as Ar neither absorb nor emit infrared radiation, as there is no net change in the dipole moment of these molecules when they vibrate. Molecular vibrations occur at energies that are of the same magnitude as the energy of the photons on infrared light. Heteronuclear diatomics such as CO or HCl absorb IR; however, these molecules are short-lived in the atmosphere owing to their reactivity and solubility. As a consequence they do not contribute significantly to the greenhouse effect.

Late 19th century scientists experimentally discovered that N_2 and O_2 did not absorb infrared radiation (called, at that time, "dark radiation") and that CO_2 and many other gases did absorb such radiation. It was recognised in the early 20th century that the greenhouse gases in the atmosphere caused the Earth's overall temperature to be higher than it would be without them.

Natural and Anthropogenic

Aside from purely human-produced synthetic halocarbons, most greenhouse gases now have both natural and anthropogenic sources. During the pre-industrial holocene, concentrations of existing gases were roughly constant. In the industrial era, human activities have added greenhouse gases to the atmosphere, primarily through the burning of fossil fuels and clearing of forests.

Ice cores provide evidence for variation in greenhouse gas concentrations over the past 800,000 years. Both CO_2 and CH_4 vary between glacial and interglacial phases, and concentrations of these gases correlate strongly with temperature. Before the ice core record, direct measurements do not exist. Various proxies and modelling suggests large variations; 500 Myr ago CO_2 levels were likely 10 times higher than now. Indeed higher CO_2 concentrations are thought to have prevailed throughout most of the Phanerozoic eon, with concentrations four to six times current concentrations during the Mesozoic era, and ten to fifteen times current concentrations during the early Palaeozoic era until the middle of the Devonian period, about 400 Mya. The spread of land plants is thought to have reduced CO_2 concentrations during the late Devonian, and plant activities as both sources and sinks of CO_2 have since been important in providing stabilising feedbacks. Earlier still, a 200-million year period of intermittent, widespread glaciation extending close to the equator (Snowball Earth) appears to have been ended suddenly, about 550 Mya, by a colossal volcanic outgassing which raised the CO_2 concentration of the atmosphere abruptly to 12%, about 350 times modern levels, causing extreme greenhouse conditions and carbonate deposition as limestone at the rate of about 1mm per day. This episode marked the close of the Precambrian eon, and was succeeded by the generally warmer conditions of the Phanerozoic, during which multicellular animal and plant life evolved. No volcanic carbon dioxide emission of comparable scale has occurred since. In the modern era, emissions to the atmosphere from volcanoes are only about 1% of emissions from human sources.

Anthropogenic Greenhouse Gases

Since about 1750 human activity has increased the concentration of carbon dioxide and of some other greenhouse gases. Natural sources of carbon dioxide are more than 20 times greater than sources due to human activity, but over periods longer than a few years natural sources are closely balanced by natural sinks such as weathering of continental rocks and photosynthesis of carbon compounds by plants and marine plankton. As a result of this balance, the atmospheric concentration of carbon dioxide remained between 260 and 280 parts per million for the 10,000 years between the end of the last glacial maximum and the start of the industrial era.

Some of the main sources of greenhouse gases due to human activity include:

— burning of fossil fuels and deforestation leading to higher carbon dioxide concentrations. Land use change (mainly deforestation in the tropics) account for up to one third of total anthropogenic CO_2 emissions.

— livestock enteric fermentation and manure management, paddy rice farming, land use and wetland changes, pipeline losses, and covered vented landfill emissions leading to higher methane atmospheric concentrations. Many of the newer style fully vented septic systems that enhance and target the fermentation process also are sources of atmospheric methane.

— use of chlorofluorocarbons (CFCs) in refrigeration systems, and use of CFCs and halons in fire suppression systems and manufacturing processes.

— agricultural activities, including the use of fertilisers, that lead to higher nitrous oxide concentrations.

The seven sources of CO_2 from fossil fuel combustion are:

1. Solid fuels (e.g. coal): 35%
2. Liquid fuels (e.g. gasoline): 36%
3. Gaseous fuels (e.g. natural gas): 20%
4. Flaring gas industrially and at wells: <1%
5. Cement production: 3%
6. Non-fuel hydrocarbons: <1%
7. The "international bunkers" of shipping and air transport not included in national inventories: 4%

The U.S. EPA ranks the major greenhouse gas contributing end-user sectors in the following order: industrial, transportation, residential, commercial and agricultural. Major sources of an individual's GHG include home heating and cooling, electricity consumption, and transportation. Corresponding conservation measures are improving home building insulation, compact fluorescent lamps and choosing energy-efficient vehicles.

Carbon dioxide, methane, nitrous oxide and three groups of fluorinated gases are the major greenhouse gases and the subject of the Kyoto Protocol, which came into force in 2005. Although CFCs are greenhouse gases, they are regulated by the Montreal Protocol, which was motivated by CFCs' contribution to ozone depletion rather than by their contribution to global warming. Note that ozone depletion has only a minor role in greenhouse warming though the two processes often are confused in the media.

Role of Water Vapor

Water vapor is a naturally occurring greenhouse gas and accounts for the largest percentage of the greenhouse effect, between 36% and 66%. Water vapor concentrations fluctuate regionally, but human activity does not directly affect water vapor concentrations except at local scales (for example, near irrigated fields). The Clausius-Clapeyron relation establishes that warmer air can hold more water vapor per unit volume. Current state-of-the-art climate models predict that increasing water vapor concentrations in warmer air will amplify the greenhouse effect created by anthropogenic greenhouse gases while maintaining nearly constant relative humidity. Thus water vapor acts as a positive feedback to the forcing provided by greenhouse gases such as CO_2.

Greenhouse Gas Emissions

Measurements from Antarctic ice cores show that just before industrial emissions started, atmospheric CO_2 levels were about 280 parts per million by volume (ppm; the units μL/L are occasionally used and are identical to parts per million by volume). From the same ice cores it appears that CO_2 concentrations stayed between 260 and 280 ppm during the preceding 10,000 years. However, because of the way air is trapped in ice and the time period represented in each ice sample analysed, these figures are long term averages not annual levels. Studies using evidence from stomata of fossilised leaves suggest greater variability, with CO_2 levels above 300 ppm during the period 7,000–10,000 years ago, though others have argued that these findings more likely reflect calibration/contamination problems rather than actual CO_2 variability.

Since the beginning of the Industrial Revolution, the concentrations of many of the greenhouse gases have increased. The concentration of CO_2 has increased by about 100 ppm (i.e., from 280 ppm to 380 ppm). The first 50 ppm increase took place in about 200 years, from the start of the Industrial Revolution to around 1973; the next 50 ppm increase took place in about 33 years, from 1973 to 2006.

Recent rates of change and emission

The sharp acceleration in CO_2 emissions since 2000 of >3% y^{-1} (>2 ppm y^{-1}) from 1.1% y^{-1} during the 1990s is attributable to the lapse of formerly declining trends in carbon intensity of both developing and developed nations. Although over 3/4 of cumulative anthropogenic CO_2 is still attributable to the developed world, China was responsible for most of global growth in emissions during this period. Localised plummeting emissions associated with the collapse of the Soviet Union have been followed by slow emissions growth in this region due to more efficient energy use, made necessary by the increasing proportion of it that is exported. In comparison, methane has not increased appreciably, and N_2O by 0.25% y^{-1}.

The direct emissions from industry have declined due to a constant improvement in energy efficiency, but also to a high penetration of electricity. If one includes indirect emissions, related to the production of electricity, CO_2 emissions from industry in Europe are roughly stabilised since 1994.

Atmospheric levels of CO_2 have set another new peak, partly a sign of the industrial rise of Asian economies led by China. Over the 2000-2010 interval China is expected to increase its carbon dioxide emissions by 600 Mt, largely because of the rapid construction of old-fashioned power plants in poorer internal provinces.

The United States emitted 16.3% more GHG in 2005 than it did in 1990. According to a preliminary estimate by the Netherlands Environmental Assessment Agency, the largest national producer of CO_2 emissions since 2006 has been China with an estimated annual production of about 6200 megatonnes. China is followed by the United States with about 5,800 megatonnes. However the per capita emission figures of China are still about one quarter of those of the US population.

Relative to 2005, China's fossil CO_2 emissions increased in 2006 by 8.7%, while in the USA, comparable CO_2 emissions decreased in 2006 by 1.4%. The agency notes that its estimates do not include some CO_2 sources of uncertain magnitude. These figures rely on national CO_2 data that do not include aviation. Although these tonnages are small compared to the CO2 in the Earth's atmosphere, they are significantly larger than pre-industrial levels.

Long-term trend

Atmospheric carbon dioxide concentration is increasing at an increasing rate. In the 1960s, the average annual increase was only 37% of what it was in 2000 through 2007. People with asthma and even active healthy adults, such as construction workers, can experience a reduction in lung function and an increase in respiratory symptoms (chest pain and coughing) when exposed to low levels of ozone during periods of moderate exertion.

Removal from the Atmosphere and Global Warming Potential

Aside from water vapor, which has a residence time of days, most greenhouse gases take many years to leave the atmosphere. Although it is not easy to know with precision how long it takes greenhouse gases to leave the atmosphere, there are estimates for the principal greenhouse gases.

Greenhouse gases can be removed from the atmosphere by various processes:

— as a consequence of a physical change (condensation and precipitation remove water vapor from the atmosphere).

— as a consequence of chemical reactions within the atmosphere. This is the case for methane. It is oxidised by reaction with naturally occurring hydroxyl radical, OH· and degraded to CO_2 and water vapor at the end of a chain of reactions (the contribution of the CO_2 from the oxidation of methane is not included in the methane Global warming potential). This also includes solution and solid phase chemistry occurring in atmospheric aerosols.

— as a consequence of a physical interchange at the interface between the atmosphere and the other compartments of the planet. An example is the mixing of atmospheric gases into the oceans at the boundary layer.

— as a consequence of a chemical change at the interface between the atmosphere and the other compartments of the planet. This is the case for CO_2, which is reduced by photosynthesis of plants, and which, after dissolving in the oceans, reacts to form carbonic acid and bicarbonate and carbonate ions.

— as a consequence of a photochemical change. Halocarbons are dissociated by UV light releasing Cl· and F· as free radicals in the stratosphere with harmful effects on ozone (halocarbons are generally too stable to disappear by chemical reaction in the atmosphere).

— as a consequence of dissociative ionisation caused by high energy cosmic rays or lightning discharges, which break molecular bonds. For example, lightning forms N anions from N_2 which then react with O_2 to form NO_2.

Stratospheric Ozone

Ozone is one of the most important trace species in the atmosphere. Ozone plays two critical roles: it removes most of the biologically harmful ultraviolet light before the light reaches the surface, and it plays an essential role in setting up the temperature structure and therefore the radiative heating/cooling balance in the atmosphere, especially the stratosphere (the region between about 10 and 60 km).

Ozone and Climatology

Ozone is mainly found in two regions of the atmosphere. Most of the ozone can be found in a layer between 10 and 60 km above the Earth's surface. This ozone region located in the stratosphere is known as the "ozone layer." Some ozone can also be found in the lower atmosphere (below 10 km) in the region known as the troposphere.

Ozone and UV

Ozone is produced by the photolysis of molecular oxygen, O_2. The oxygen atom, O, produced by this photolysis recombines with O_2 to form ozone, O_3. Ozone formation primarily occurs in the tropical upper stratosphere, where it is transported poleward and downward by the largescale Brewer-Dobson circulation. The formation of ozone by the photolysis of molecular oxygen removes most of the incident sunlight with wavelengths shorter than 200 nm. The wavelengths between 200 and 310 nm are removed by the photolysis of ozone itself. This photolysis of ozone in the stratosphere is the process by which most of the biologically damaging ultraviolet sunlight (UV-B) is filtered out. As this filtering process occurs, the stratosphere is heated. This heating is responsible for the temperature structure of the stratosphere, where the temperature increases as the altitude increases. Without this filtering, larger amounts of UV-B would reach the surface.

Ozone and Climate Change

If ozone in the stratosphere were to be removed, the stratosphere would cool. How a cooler stratosphere affects radiative balance in the rest of the atmosphere has been the subject of many detailed studies. These studies have been reanalysed and integrated into the latest Intergovernmental Panel on Climate Change (IPCC) report, "Climate Change 1994: Radiative Forcing of Climate Change". The conclusion of that report is that stratospheric ozone loss leads to a "small but non-negligible offset to the total greenhouse forcing from CO_2, N_2O, CH_4, CFCs...." It is ironic that the size of the negative radiative forcing from ozone loss is nearly equal to the positive radiative forcing from chlorofluorocarbons (CFCs), the source of the stratospheric ozone loss. The size of the radiative forcing due to stratospheric ozone loss has also been shown to be very sensitive to the profile shape assumed for that loss.

While the global amount of ozone is fairly constant, there are significant local, seasonal, and long-term changes. The seasonal ozone changes are basically determined by the winter-summer changes in the stratospheric circulation. Since ozone has a lifetime of weeks to months in the lower stratosphere, the amount of ozone can strongly vary due to transport by stratospheric wind systems. Since weather conditions in the stratosphere, as in the troposphere, vary from year to year, there is also interannual variability in ozone amounts. Interseasonal changes in ozone are also linked to the 11-year solar cycle in UV output and the amount of volcanic aerosols in the stratosphere. Changes in ozone have also been linked to anthropogenic pollutants, especially the release of manmade chemicals containing chlorine. In the section below we describe the more-significant recent global changes in ozone observed by a variety of instruments.

Polar ozone changes

The first ozone measurements in the Antarctic were made during the 1950s. A Dobson instrument was installed at Halley Bay in late 1956 in preparation for the International Geophysical Year in 1957. One of the first discoveries made by this instrument was that the seasonal cycle of ozone in the south polar region is very different from that which had been observed in the north. This was noted in a review article by Dobson, which pointed out that its cause was a difference in the circulation patterns of the Antarctic relative to the Arctic. In the Arctic, the total ozone amount grew rapidly in the late winter and early spring to about 500 Dobson Units (DU). (A DU is one milliatmosphere-cm of pure ozone; a layer of pure ozone that would be 0.001-cm thick under conditions of Standard Temperature and Pressure

[STP].) In contrast, the Antarctic early springtime amounts remained near 300 DU.

The Dobson instrument at Halley Bay continued to make measurements each year. Farman et al. showed that the springtime ozone amounts over Halley Bay had declined from nearly 300 DU in the early 1960s to about 180 DU in the early-to-mid 1980s. This result has been confirmed at a number of other stations and shown using satellite data to occur over an area larger than the Antarctic continent. These large ozone changes implied that losses must be taking place in the lower stratosphere where most of the ozone exists. This was shown to be true in a series of ozonesonde measurements. More-recent sonde measurements have shown instances of near-zero concentrations of ozone over a 5-km altitude range. Aircraft measurements and satellite measurements confirm and show further details of these ozone changes.

The 1996-1997 Northern Hemisphere winter experienced a significant ozone depletion over the Arctic and subsequent total ozone values achieved record low values in the spring. Long term records of the Total Ozone Mapping Spectrometer (TOMS) and ground based observations show a downward change over the past several years occurring mostly in February and March and confined to the lower stratosphere similar to the depletion over the Antarctic. Chlorine radicals have been conclusively identified as the causes of ozone depletion now in both hemispheres. Measurements of ClO by the UARS MLS instrument observed elevated levels in late February over Northern Hemisphere polar regions. The winter of 1996-1997, showed extremely low temperatures in the Stratosphere. These cold temperatures led to the formation of polar stratospheric clouds whose particles shift chlorine gas away from its HCl reservoir to active ClO through heterogeneous chemistry.

This is the primary mechanism producing the Antarctic ozone hole. Although the buildup of chlorine has occurred approximately uniformly in both hemisphere the unusually low temperatures reached in high northern latitudes mostly likely precipitated the concurrent ozone losses over the Arctic.

Midlatitude ozone loss

Midlatitude ozone loss estimates must be extracted from long time series using statistical models. The longest time series of total ozone data is from

Arosa in Switzerland. Changes in the profile of ozone with altitude can be deduced from sonde data or from the Stratospheric Aerosol and Gas Experiment (SAGE) satellite measurements. Analyses of sonde data show ozone decreases between the tropopause and about 24-km altitude. Analyses of SAGE data show larger decreases than those derived from sondes. SAGE results show negative ozone trends in the lower stratosphere in the tropics. Column ozone changes deduced from SBUV and TOMS show only small downward trends. Hollandsworth et al. used SBUV profile and total ozone trends to deduce that ozone in the tropics below 32 hPa has increased slightly over the last decade. The resolution of the uncertainty in the magnitude of lower stratospheric and upper tropospheric ozone trends is an important measurement and analysis issue for the coming years.

Stratospheric Ozone Distribution

Chemical processes

Ozone is being continuously created and destroyed by the action of ultraviolet sunlight. The overall amount of ozone in the global stratosphere is determined by the magnitude of the production and loss processes and by the rate at which air is transported from regions of net production to those of net loss. Production of ozone requires the breaking of an O_2 bond, with the extra or "odd" oxygen atom attaching to another O_2 to form O_3. This most frequently occurs via the photodissociation of O_2 by solar ultraviolet radiation. In the lower stratosphere and troposphere ozone can also be produced by photochemical smog-like reactions. In these reactions H or CH_3 or higher hydrocarbon radicals attach to an O_2 forming HO_2 or CH_3O_2, etc., which then react with NO. This reaction breaks the O_2 bond by forming NO_2 (which is really ONO). When NO_2 is photolyzed an O atom is formed which reacts with O_2 to form O_3.

Loss of ozone occurs when an O atom reacts with O_3 to re-form the O_2 bond. More importantly, this loss process is catalyzed by the oxides of hydrogen, nitrogen, chlorine, and bromine. These oxides are produced in the stratosphere from longlived, unreactive molecules released at the surface of the Earth. The major source molecules for HO_x (HO_x is chemical shorthand for the sum of all the hydrogen radicals—OH and HO_2, mostly) are methane (CH_4) and water vapor (H_2O). The main source of NO_x (NO_x is chemical shorthand for the sum of all nitrogen radicals NO, NO_2, N_2O_5, NO_3, mostly) is nitrous oxide (N_2O). The major sources of chlorine are industrially- produced CFCs and naturallyoccurring methyl chloride (CH_3Cl). The major sources of bromine are methyl bromide (CH_3Br) and the halons (CF_3Br and CF_2ClBr). These source molecules are transported to the stratosphere where they react or are photodissociated to produce the catalytically-active oxide radicals.

The catalytic efficiency of hydrogen, nitrogen, chlorine, and bromine oxides is determined by a set of interlocking reactions which convert the active oxides to catalytically-inactive temporary reservoirs, such as HNO_3, HCl, $ClONO_2$, H_2O, HOCl, HOBr, and $BrONO_2$, and vice versa. In the lower stratosphere, the balance between catalytic oxides and temporary reservoirs is strongly affected by reactions on the surfaces of stratospheric aerosols. The balance is even more profoundly affected in the polar winter by reactions on the surface of Polar Stratospheric Cloud (PSC) particles. In the early spring, the chlorine balance is shifted to almost 100% ClO_x (ClO_x is shorthand for the sum of all chlorine radicals, C1O, Cl_2O_2, Cl). This shift in the chemical balance results in a large calculated chemical sensitivity of ozone towards chlorine perturbations and a relatively small calculated sensitivity of ozone towards nitrogen oxide perturbations.

Although the basic outline of the chemistry controlling stratospheric ozone is now known, many important aspects of the problem remain to be solved. The primary difference between the Northern and Southern Hemispheric polar ozone loss regions appears to be a result of the "denitrification" that occurs in the Antarctic winter. Denitrification means the removal of nitrogen oxides and HNO_3 by large particles which fall into the troposphere. Denitrification takes place when temperatures are cold enough to form large stratospheric ice crystals. When springtime comes there are no nitrogen oxides to convert ClO_x to $ClONO_2$ and slow down the rate of ozone depletion. There is some evidence for denitrification when temperatures are not cold enough to form ice crystals. Under those conditions the mechanism for denitrification is not completely understood.

Transport

Much of the currently-observed ozone interannual variability in the stratosphere is controlled by dynamical processes. In particular, this variability is driven by such processes as the quasi-biennial oscillation (QBO), El Niño-Southern Oscillation, tropospheric weather systems which extend into the stratosphere, and long-term fluctuations in planetary wave activity. The annual cycle of total ozone is largely driven by transport effects. These low tropical ozone values occur in spite of the large ozone production rates in the tropics. If ozone production were precisely balanced by ozone loss everywhere, total ozone would have extremely high values in the tropics. The observed tropical low values result from vertical advection of low-ozone air from the tropical troposphere into the tropical stratosphere, and the subsequent transport of this air poleward and downward into the extratropics and polar regions. This advective circulation is known as the Brewer-Dobson circulation.

The redistribution of ozone from the production region at low latitudes to extratropical latitudes is modulated by a variety of processes. Foremost among these processes is the annual cycle in the circulation. It is now recognised that the Brewer-Dobson circulation is primarily controlled by large-scale waves in the winter stratosphere. As these waves propagate through the westerly winds that dominate the winter stratosphere, they exert a westward zonal drag, which through the Coriolis force leads to a poleward and downward transport circulation, which in turn drives the temperatures away from radiative equilibrium. The large-scale waves breaking in the winter upper stratosphere also produce lifting in the tropics. Since the lifetime of ozone increases with pressure, the poleward downward circulation causes ozone to accumulate in the lower stratosphere over the course of the winter. Since the large-scale waves are not present in the summer, the poleward and downward circulation is significantly weakened, and ozone amounts which have built up during winter begin to decrease due both to transport into the troposphere and to photochemistry.

The exchange of mass between the troposphere and the stratosphere is the focus of considerable current research. Stratosphere-troposphere exchange is important for the budget of ozone in the lower stratosphere as well as the ozone budget in the troposphere. Upward transport occurs in the tropics, but the exact mechanism controlling the transport is not clear. Current research is focussing on the role of subvisible cirrus and the radiative impact of infrared (IR) heating of subvisible cirrus. Downward transport (stratosphere to troposphere) takes place in midlatitudes through jet stream folds—but the frequency and amount of mass irreversibly moving through these folds is still not understood.

Aerosols and Polar Stratospheric Clouds (PSCs)

It is now known that knowledge of stratospheric aerosols and PSCs is very important to our understanding of stratospheric ozone. The surfaces of aerosols and PSCs are sites for heterogeneous reactions which can convert chlorine from reservoir to radical forms. Likewise, radical nitro-gen forms can be sequestered as nitric acid to shift the chemical loss process.

Aerosols

The long-term stratospheric aerosol record reveals at least three components: episodic volcanic enhancements, PSCs and clouds just above the tropical tropopause, and a background aerosol level. At normal stratospheric temperatures, aerosols are most likely super-cooled solution droplets of $H_2SO_4 \cdot H_2O$, with an acid weight fraction of 55 to 80%. The primary source of stratospheric aerosols is volcanic eruptions that are strong enough to inject SO_2 buoyantly into the stratosphere. Aerosol sizes range from hundredths of a micrometer to several micrometers. Although there is some variability, especially just after a volcanic eruption, a log-normal size distribution of spherical particles appears to aptly describe the aerosol. Just after an eruption, the size distribution becomes bimodal, and some particles are nonspherical because of the addition of crustal material. After an eruption, the SO_2 is converted to H_2SO_4, which condenses to form stratospheric sulfuric acid aerosols, with a time scale of about 30 days. Subsequently, aerosol loading decreases due to a combination of sedimentation, subsidence, and exchange through tropopause folds. The loading decreases with an e-folding time of 9-to-12 months, although this appears quite variable with altitude and latitude.

The net effect of this post-volcanic dispersion and natural cleansing is a greatly enhanced aerosol concentration in the upper troposphere after a major eruption, especially poleward of about 30° latitude. Except immediately after an eruption, stratospheric aerosol droplets tend to be concentrated into 3 distinct latitudinal bands—one over the equatorial region (to 30°) and the other over each high-latitude region, 50° to 90° N and S. Following a low-latitude eruption, aerosol is dispersed into both hemispheres, whereas following a mid-to-high-latitude eruption, aerosols tend to stay primarily in the hemisphere of the eruption. Potential sources of a background aerosol component include carbonyl sulfide (OCS) from the oceans, low-level SO_2 emissions from volcanoes, and various anthropogenic sources, including industrial and aircraft emissions. Also, it is not clear whether there is an upward trend in this background aerosol, as has been hypothesized and linked to increasing aircraft emissions, since any increase may be due to incomplete removal of past volcanic aerosol.

Stratospheric aerosol loading in 1979 was approximately 0.5×10^{12}g (0.5 Mt), thought to be representative of background aerosol conditions. The present status of the aerosol is one of enhancement due to the June 1991 eruption of Pinatubo (15.1° N, 120.4° E), which produced on the order of 30×10^{12}g (30 Mt) of new aerosol in the stratosphere, about 3 times that of the 1982 eruption of El Chichón. This perturbation appears to be the largest of the century, perhaps the largest since the 1883 eruption of Krakatoa. By early 1993, stratospheric loading decreased to approximately 13 Mt, about equal to the peak loading values after El Chichón. Measurements in 1995 showed that the aerosol levels were approaching background levels.

Polar Stratospheric Clouds (PSCs)

The interannual variability in PSC sightings has been addressed by Poole and Pitts, who analysed more than a decade of data from the spaceborne Stratospheric Aerosol Measurement (SAM) II sensor. They found noticeable variability in PSC sightings in the Antarctic from year to year, even though the southern polar vortex is typically quite stable and long-lived. This variability was found to occur late in the season and can be explained qualitatively by temperature differences. Poole and Pitts found even more year-to-year variability in SAM II Arctic PSC sighting probabilities. This was expected since the characteristics and longevity of the northern polar vortex vary greatly from one year to the next. The year-to-year variability in Arctic sighting probabilities can also be explained qualitatively by differences in temperature, e.g., zonal mean lower stratospheric temperatures in February 1988 were as much as 20 K colder than those one year earlier.

Solar ultraviolet and energetic particles

Since ozone formation is fundamentally linked to the levels of ultraviolet radiation reaching the Earth, natural variations in that radiation must be understood in order to detect trends. The ultraviolet comprises only one-to-two percent of the total solar radiation, but it displays considerably more variation than the longer wavelength visible radiation. For example, from 1986 to 1990 the solar UV increased with onset of the 11-year solar cycle and resulted in an increase of global total ozone of almost 2%. This natural increase in ozone is comparable to the suspected anthropogenic decrease and needs to be understood in order to totally separate the anthropogenic decrease from this natural change. Studies of total ozone trends typically subtract solar cycle and other natural changes from the total ozone record in trend resolution. Thus, more-quantitative knowledge of this natural solar-cycle-induced total ozone change would be especially valuable.

Changes in energetic particle flux from the sun penetrate into the middle atmosphere and may also drive the natural ozone variations. A series of solar flares in 1989 spewed solar particles into the Earth's polar cap regions (greater than 60° geomagnetic latitude) and led to polar ozone depletion. Further studies related to the very large solar particle events (SPEs) of October 1989 have predicted ozone depletions lasting for several months after the SPEs. Although SPEs of this magnitude occur infrequently (only two have been observed in the past 25 years), they need to be understood more completely to be able to separate natural from anthropogenic ozone effects.

Relativistic electron precipitations (REPs) have been predicted to contribute substantially to the odd nitrogen budget of the stratosphere and, therefore, have been predicted to play a large role in controlling ozone in this region. Another investigation has failed to find any REP-caused ozone depletion. Further work determined that REPs in May 1992, the largest measured relativistic electron flux precipitating in the atmosphere between October 1991 and July 1994, added only about 0.5 to 1% of the global annual source of odd nitrogen to the stratosphere and mesosphere. The actual importance of REPs in regulating ozone is thus not well understood nor characterised, and further work on REPs is required to thoroughly determine their importance regarding modulation of stratospheric ozone.

Modeling the Ozone Distribution, Assessments

Models of the stratosphere provide the only means to attempt quantitative prediction of global change, or to evaluate the impact of natural or anthropogenic changes in composition on the stratospheric ozone and climate. In addition, models provide a means to integrate observations and theory, to provide tests of mechanisms for chemical, dynamical, or radiative changes, and to enable interpretation of observations from different platforms.

Two-dimensional models

Two-dimensional (2-D) models are used by several research groups. The models predict the behavior of ozone and other trace gases in reasonable agreement with measurements. Because of these favorable comparisons to measurements, these models have been utilised recently in many atmospheric studies, for example: 1) the response of the middle atmosphere due to solar variability was studied by Brasseur, Huang and Brasseur, and Fleming et al.; 2) the influence of the Mt. Pinatubo eruption on the stratosphere was studied by Kinnison et al. and Tie et al.; and 3) the effects of proposed stratospheric aircraft on atmospheric constituents were studied by Pitari et al., Weisenstein et al., Considine et al., and Considine et al.

These 2-D models have also been used to produce multi-year simulations of the response of stratospheric ozone to perturbations of the source gases such as CFCs from which chlorine radicals are produced. An outstanding issue regarding simulations of the stratospheric ozone response to chlorine increases is the lack of ability of 2-D models to accurately predict the ozone trend in the middle and high northern latitudes over the 1980-to-1990 time period. Since the 2-D models predict a smaller trend than observed, it is believed that the models do not adequately model all of the relevant processes and thus require further development.

Three-dimensional models

The three-dimensional (3-D) (or general circulation) model with full interaction between chemical, dynamical, and radiative processes remains elusive. The present generation of general circulation models (GCMs) generates unrealistic temperature fields which, in turn, alter the photochemistry. The unrealistic temperatures are related to problems with the model transport circulation. For example, the polar regions are persistently cold in GCMs, which suggests that there is insufficient adiabatic heating (or descent) in the winter polar region. Correspondingly, there will be insufficient ascent in the tropics, which weakens the transport from the troposphere into the stratosphere.

Subtle changes in the general circulation of the atmosphere in 3-D models can alter and distort the chemical feedbacks. For example, Rasch et al. report on a two-year simulation using version 2 of the National Center for Atmospheric Research (NCAR) Middle Atmosphere Community Climate Model (MACCM2). A chemical scheme for 24 reactive species, or families, is run as part of this simulation. This model is partially coupled in that the water vapor predicted by MACCM2 is connected to the chemical source of water through oxidation of methane. Prescribed ozone is used in the radiative calculation. In this simulation, the calculated upper stratospheric ozone is substantially lower than is observed; much of the difference is attributed to the lower CH_4, compared to observations by the UARS Halogen Occultation Experiment (HALOE). This bias leads to excessive ClO and excessive destruction of O_3. In effect, the error in this longlived trace gas, which results from the weak transport circulation, leads to noticeable errors in ozone.

The difficulties described above show why most 3-D modeling efforts have focused on "off-line" calculations, i.e., use of chemistry and transport models (CTMs) in which the wind and temperature fields are input from a GCM or from a data assimilation assimilation system. For either approach, there are computational advantages, as the same set of winds and temperatures is used many times. Furthermore, the effects of modifications to the chemical scheme can be isolated, and their effects understood, without the complications caused by feedback processes. A further advantage of the use of assimilated winds and temperatures is that the results of constituent simulations may be compared directly with observations with no temperature biases such as those found in GCMs. This is particularly important for the study of processes which have a temperature threshold, such as heterogeneous reactions on PSC surfaces. The most information is gleaned when the model is sampled in a manner consistent with the satellite sampling.

The "off-line" approach has been used successfully for many years and is used to test chemical and transport mechanisms, as well as to interpret observations. These tests include:

1) assessment of the importance of transport of air with high levels of reactive chlorine to middle latitudes;
2) assessment of the rate of ozone loss within the Northern Hemisphere vortex and identification of the variables to which the calculation is sensitive;
3) determination of the importance of upper tropospheric synoptic-scale systems on the vortex temperature, as well as their influence on the transport and mixing of air which has experienced temperatures cold enough for PSC formation; and
4) examination of the impact of ozone transport following the breakup of the Antarctic polar vortex on the global ozone budget.

These 3-D studies provide a picture of the important physical processes which control polar ozone loss. However, because of computer resource restrictions, it is not yet possible to make full 3-D model long-range predictions, including possible influence of the ozone loss on lower stratospheric temperature and climate. For example, future temperature changes may have a significant impact on the Northern Hemisphere vortex. The full 3-D model, with all relevant chemical, dynamical, and radiative processes and feedbacks among them, has yet to be developed.

Major Scientific Issues and Measurements

Changes in the ozone layer can be divided into two categories: natural changes and man-made changes. Separating these components is the goal of much ozone and trace gas research. Since ozone can be transported by stratospheric winds, there is significant interannual variability in column ozone amounts. Ozone is likewise influenced by aerosol amounts (through heterogeneous chemistry), the formation of nitrogen radicals associated with high-energy particles, and variations in the ultraviolet radiation from the sun. Manmade changes generally include increased chlorine and hydrogen amounts from industrial gases and increased aerosols and nitrogen radicals from airplane exhaust. Many of our current scientific issues and future measurement needs center around the interaction of the ozone layer with these pollutants and separating natural changes in the ozone layer from man-made processes.

Natural Changes

Interannual and long-term variability of the stratospheric circulation

Because the stratospheric circulation is strongly dependent on the dissipation of large-scale waves in the stratosphere, interannual variability of the wave amplitudes has an important impact on ozone transport. Winds and temperatures derived from 3-D GCMs and assimilation models include such interannual variability and can be used to assess the impact on ozone transport. 2-D models can incorporate prescribed variability to simulate interannual ozone transport.

Accurate assessment of the large-scale waves and the transport circulation is necessary for understanding the variability of ozone trends. One of the failures of the 3-D models is inadequate simulation of the QBO. The QBO is a 24-30-month oscillation of the zonal wind in the tropical lower stratosphere that is driven by tropical waves. The QBO affects the stratospheric temperature distribution and produces a secondary circulation which transports trace gases and aerosols. For ozone, the QBO can generate variations from the climatological mean of 5-10 DU in the tropics. There is also a QBO-ozone signal outside the tropics of 10-20 DU.

The QBO provides one of the largest components of the interannual variability of the column ozone values. Because the geostrophic relationship breaks down in the tropics, direct tropical wind measurements are critical to precisely measuring the QBO and for understanding the effects of the QBO on the circulation. Data-sparse regions and infrequent sampling of wind fields all preclude good quantitative studies of the tropical circulation and its effect on ozone.

External influences (solar and energetic particle effects)

Solar ultraviolet radiation and precipitating energetic particles can strongly influence ozone amounts. In order to understand the anthropogenic changes in ozone, we must maintain reliable measurements of the solar ultraviolet input to the middle atmosphere. Solar variations in the UV produce ozone changes on the same order of magnitude as the current observed midlatitude changes. Proxies for the UV changes have been historically used to estimate the response of ozone to solar ultraviolet changes. With direct measurements from UARS, these proxies have been shown to inadequately represent changes in ultraviolet flux. Particle events generate NOx compounds which catalytically destroy ozone, but these events tend to be confined to the upper stratosphere. Large events, which tend to be more episodic, may affect polar ozone at lower levels. The impact of NOx generation through particle precipitation on the natural ozone layer is a major scientific question.

Natural aerosols and PSCs

Irregular volcanic inputs of SO_2 with the subsequent formation of sulfate aerosols have an impact on the ozone layer. There is some evidence suggesting that increasing amounts of background aerosols are a result of subsonic aircraft emissions in the lower stratosphere. A major scientific question is whether the background amounts of these aerosols are increasing and, if so, determining their origin. Monitoring the aerosol amounts within the stratosphere and determining their trend is a primary measurement requirement to understand ozone loss. During the 1980s it became apparent that aerosols play an important role in the chemistry of the stratosphere.

Observations of large decreases in ozone over Antarctica during the Southern Hemisphere spring were not accounted for by theory, until several researchers hypothesized that heterogeneous reactions on PSCs might be converting inactive chlorine compounds into reactive forms. In a similar fashion to PSCs, heterogeneous reactions upon sulfuric acid droplets at midlatitudes convert N_2O_5 into HNO_3 and shift the ratio of HNO_3 to NO_2 normally present in the stratosphere. Throughout the stratosphere, reactions on and inside aerosol particles are therefore important.

To understand the effectiveness of the heterogeneous (gas phase/aerosol phase) reactions, it is important to know:

a) the temperatures of the aerosol particles;

b) the surface and volume densities of the aerosol particles, which are derived from the aerosol extinction, and a knowledge of the size distribution;

c) the composition (the mixing ratios of H_2O, H_2SO_4, and HNO_3 in ppbv) and phase (liquid/solid/amorphous solid solution) of the aerosol particles;

d) the concentration of the reactants in the aerosol (e.g., the concentration of HCl); and

e) the duration of time over which the heterogeneous reactions occur.

A theoretical framework, by which heterogeneous rates of reaction are quantified, is given in Hanson et al. An important research goal is the ability to observe the yearly episodes of ozone loss in the polar regions (e.g., the Antarctic ozone hole), to measure this loss as reservoir chlorine levels change with time, and to be able to relate the changes in observed ozone to a quantitative understanding of heterogeneous processes.

In principle, one should be able to identify the composition of stratospheric aerosol from multi-wavelength extinction data. Multi-wavelength observations of midlatitude sulfuric acid droplets have an extensive history. Observations of El Chichón aerosol, post El Chichón aerosol, and of Mt. Pinatubo aerosol yield spectral data consistent with theoretical expectation. Several years ago, ice and nitric acid trihydrate (NAT) particles were thought to be the primary composition of PSCs. Recent studies have shown that some PSC particles are liquid (the ternary solution of $HNO_3/H_2O/H_2SO_4$), and not that of crystalline NAT. As additional laboratory coldtemperature measurements of the indices of refraction of PSC composition candidates become available, the ability to classify PSC composition from spectra will improve.

Although PSCs are now known to be instrumental in polar ozone loss, their amounts and types must be monitored. The major difference between the Antarctic ozone depletion and the less-severe Arctic depletion appears to be the result of a lack of denitrification in the Arctic. Fundamentally, denitrification is a function of temperature and the size of PSCs. Above frost point the PSC size is generally too small to precipitate nitric acid from the stratosphere. If temperatures reach frost point, larger PSCs form, which are able to remove nitrogen acid from the lower stratosphere. The temperature history of the air parcel may play an important role in the PSC size distribution as well. Photolysis of the nitric acid is key to halting the ozone depletion during winter.

With the increase of greenhouse gases, the stratosphere is expected to cool and thus increase the probability of PSC formation as well as increase the surface area and heterogeneous reaction rates on sulfate aerosols. Preliminary studies suggest greenhouse gas increase could have a major role in polar ozone depletion through increased probability of PSC formation. Monitoring stratospheric aerosol loading and PSC amounts is critical for understanding ozone loss.

Man-made Changes

Man-made changes in ozone mostly arise from the manufacture of unreactive chlorine-containing compounds such as the CFCs (chlorine source gases). These compounds reach stratospheric altitudes where photolysis by ultraviolet radiation releases chlorine with subsequent destruction of ozone through catalytic cycles. Aviation also has an impact on ozone through the release of nitrogen radicals in aircraft exhaust. Both of these anthropogenic effects are discussed below.

Trends in chlorine source gases

All chlorine in the stratosphere comes from tropospheric sources, predominantly the man-made CFCs and chlorocarbons. The man-made sources account for about $7/8^{th}$ of the total stratospheric chlorine. CFCs are cur-rently being phased out in favor of the hydrochlorofluorocarbons (HCFCs). Extensive measurements of the chlorofluorocarbons CFC-11 (CCl_3F), CFC-12 (CCl_2F_2), and CFC-113 (CCl_2FCClF_2) have indicated a steady increase in their tropospheric mixing ratios for more than a decade. Most recent data suggest that the growth rate for these species has begun to decrease. Measurements taken from Tasmania suggest that levels of the important chlorocarbon CCl_4 in the troposphere are also decreasing.

As HCFCs are introduced as substitutes for CFCs, it may be expected that their mixing ratios in the troposphere will increase well into the next century. HCFC-22 (CHClF2) data show a near-linear growth rate in recent years. HCFC-141b and HCFC-142b have been available only recently as CFC replacements. These species are clearly increasing in the troposphere, but further data is required to get reliable growth rates for long-term studies. CH_3CCl_3 data also indicate a reduced growth rate that is a result of recently-reduced emissions, but also possibly due in part to increasing hydroxyl (OH) levels. Data for dichloromethane (CH_2Cl_2), methyl chloride (CH_3Cl), and chloroform ($CHCl_3$) currently exhibit no long-term trends. Continued tropospheric measurements of these gases are required to estimate ozone depletion potential.

Stratospheric chlorine

An extensive compilation of measurements of chlorine source gases in the stratosphere can be found in Fraser et al. The most comprehensive suites of simultaneous measurements of chlorine constituents in the stratosphere include the Atmospheric Trace Molecule Spectroscopy (ATMOS) experiments of 1985, 1992, and 1993, and the Airborne Arctic Stratospheric Expedition II (AASE II) measurements of 1991, 1992. The data from these missions have provided invaluable information on the stratospheric chlorine burden and the partitioning among the various chlorine species.

Based upon the 1985 ATMOS data, Zander et al. determined a total stratospheric chlorine level of 2.55 ± 0.28 ppbv. Further, they concluded that above 50 km most of the inorganic chlorine was in the form of hydrogen chloride (HCl), and that the partitioning of the chlorine among sources, sinks, and reservoir species was consistent with that level of total chlorine. From the 1992 ATMOS flights, total stratospheric chlorine was estimated to be 3.4 ± 0.3 ppbv, an increase of approximately 35% in seven years. This increase is consistent with that predicted by models. Schauffler et al. inferred total chlorine levels levels of 3.50 ± 0.06 ppbv from the AASE II data near the tropopause, a value which is in excellent agreement with the 1992 ATMOS values.

Recent HCl data (55 km) from HALOE on UARS reveal a trend in HCl versus time at 55 km (v18) compared with the estimated total Cl trend based on tropospheric emissions. Of the total stratospheric burden, only about 0.5 ppbv is estimated to arise from natural sources in the troposphere, but these estimates have yet to be confirmed by direct or remote observations. HCl emissions from major volcanic eruptions provided negligible perturbations to the levels of HCl in the stratosphere.

Depletion of ozone by stratospheric chlorine

Estimates of the severity of ozone depletion in the future can only be determined by atmospheric model simulations. The level of confidence in these models is based upon their ability to simulate present atmospheric distributions and their ability to simulate recent (decadal) trends. Model simulations of ozone change spanning the period 1980 to 2050 were conducted as part of the WMO assessment process. Two scenarios were adopted for the assessment studies: 1) the emissions of halocarbons follow the guidelines in the Amendments to the Montreal Protocol, Scenario I; and 2) partial compliance with the guidelines, Scenario II.

Decreases of up to approximately 6.5% are seen to occur just prior to 2000. The recovery time to 1980 levels varies widely for the different models, from as early as 2020 to well past 2050. The individual models all showed reasonable agreement among themselves for the present-day ozone distributions, but begin to differ substantially as the atmosphere is perturbed away from its existing state by increasing levels of nitrous oxide, methane, halocarbons, and other influences.

Uncertainties in the absolute levels of depletion predicted by the models are difficult to evaluate for these long-term scenario calculations. The trends in the source gases are changing, and the trends in the stratospheric reservoir gases, which are dependent on transport into the stratosphere, will respond. Thus, measurements of the chlorine source and stratospheric reservoir gases must be made to test models against observations. Critical gases in the suite of required measurements are the reservoirs HCl and $ClONO_2$. The predictive capability of these assessment models directly rests on additional measurements of chlorine source gases, reservoir gases, and gases which are sensitive to transport processes.

Effects of aircraft exhaust

Long-lived source gases (e.g., N_2O, CH_4) are unreactive in the troposphere and hence can enter the stratosphere at the ambient tropospheric concentrations. In the stratosphere, these gases undergo photolysis or react with radicals to release their potential ozone-destroying catalytic agents. In contrast, aircraft flying in the stratosphere will directly inject catalytic agents into the stratosphere. The primary agents for potential ozone change which have been considered in studies of aircraft exhaust are the nitrogen oxides (NO_x) and water vapor (which leads to HO_x). Now that heterogeneous reactions on background aerosols and PSCs are known to play an important role in the ozone balance of the stratosphere, the evaluation of the effects on ozone of NO_x from supersonic aircraft flying in the stratosphere has changed significantly.

The impact on column ozone of a fleet of supersonic transports (now referred to as High Speed Civil Transports [HSCTs]) is now calculated to be of the order of 1% or less. An important possibility is that the sulfur in the exhaust will lead to the generation of numerous small particles which will add to the aerosol surface area. An increase in surface area will enhance the conversion of chlorine from its reservoirs to ClOx and thus could lead to an increased loss rate for ozone. Another possibility is that the other condensibles in the exhaust, water vapor, and nitric acid (from NO_x) could impact the formation or duration of PSCs. Initial calculations show this effect to be small, and transport studies show that injection into the polar vortex is unlikely, but there is still uncertainty about what will happen as the stratosphere cools with increasing CO_2 concentrations.

All of the chemical effects of HSCT exhaust depend on how much of the exhaust products accumulate in the stratosphere and where they accumulate. The same is true for the exhaust of the subsonic fleet, which is released in the upper troposphere and lower stratosphere. The three major potential effects of the subsonic fleet of aircraft are ozone increase due to the smog-like photochemistry of NO_x, CO_2 increase due to fuel consumption, and cirrus cloud formation from the water vapor.

The importance of aircraft NO_x to ozone generation in the upper troposphere and lower stratosphere is not completely understood. Aircraft NO_x sources have to be compared to the NO_x sources due to lightning, stratospheric intrusions, and the lofting of ground-level pollution in cumulus clouds. Thus the role of aircraft as a source of upper tropospheric NO_x and its impact on lower stratospheric ozone is uncertain. Also uncertain is whether heterogeneous chemistry on ice crystals plays a significant role in the NO_x budget.Understanding the impact of supersonic and subsonic aircraft exhaust on the stratospheric chemical balance is a complex problem. Knowledge of meteorological conditions is required to compute exhaust dispersion. Knowledge of aerosol chemistry is required to understand the aerosol formation process (from sulfur in fuels) and its impact on the background conditions. Finally, a good understanding of the lower stratosphere chemistry is required to understand the direct impact of the NO_x pollutants.

Required Measurements

Meteorological Requirements

An understanding of the photochemistry of the stratosphere is clearly contingent on high-quality observations of temperature. The temperature field affects stratospheric physical processes in a number of ways. First, temperature fields are used to calculate geopotential heights and winds via the hydrostatic and geostrophic approximations. Second, temperatures affect the radiation field, particularly in relation to the longwave cooling in the stratosphere. Third, temperatures affect the chemistry via temperaturedependent reaction rates, and via the formation of PSCs (the indirect cause of the ozone hole). Hence, accurate and precise temperatures provide a basic foundation for stratospheric chemistry, radiation, dynamics, and transport.

The National Plan for Stratospheric Monitoring 1988-1997 set down the minimum requirements for meteorological variables between 1000 and 0.1 hPa. Their requirements were: 2.7-km vertical resolution; 12-hour time resolution; 1-K precision for temperature, and 5 m s-1 precision for winds. Current radiosonde and rawinsonde measurements have precisions of a few tenths of a kelvin and 1-4 m s-1 wind speed in precision. Unfortunately, the balloon-borne rawinsonde system is limited to altitudes below 30 km. For higher altitudes, the meteorological rocket network provided some data, but the network has been effectively discontinued. Satellite systems are now relied upon to provide all of the meteorological information above 30 km.

The National Oceanographic and Atmospheric Administration (NOAA) TIROS Operational Vertical Sounder (TOVS) (Microwave Sounding Unit [MSU], Stratospheric Sounding Unit [SSU], and High-Resolution Infrared Sounder [HIRS]) SSU instrument has an error of 2 K at 10 hPa rising to 4 K at 1 hPa. The TOVS weighting functions are about 10-12-km deep. Later

NOAA sounders use the Advanced Microwave Sounding Unit (AMSU) instead of MSU/SSU. The AMSU weighting functions are about half the depth of the TOVS functions. AMSU temperature measurements are limited to the atmosphere below 50 hPa.

The UARS Microwave Limb Sounder (MLS) has a vertical resolution of a few km, although its horizontal coverage is inferior to nadir-sounding TOVS and AMSU instruments. Improved understanding of stratospheric chemistry and heterogeneous processing suggests that improvement of temperature measurements will have an impact on our ability to predict where the heterogeneous reactions will take place.

Lower stratospheric temperature measurements made during the numerous polar aircraft missions suggest that the meteorological analyses (based upon TOVS) in the Southern Hemisphere are warm biased by about 2 K. This suggests that the Earth EOS stratospheric temperature accuracy requirements should be less than 0.5 K. It is also important that good temperature measurements be made near the tropical tropopause, especially in cloudy regions where air is entering the stratosphere through the tropopause.

Direct stratospheric wind data are needed where the divergence fields are significant (e.g., the tropics). The current wind requirement for assimilation models is an unbiased horizontal wind field accuracy of 2-5 m s-1. These should be global measurements with a vertical resolution of a few kilometers. Presently, the UARS High Resolution Doppler Imager (HRDI) satellite wind instrument makes these measurements at the upper end of the limit.

A joint effort between NOAA, the Department of Defense, and NASA will produce the National Polar Orbiting Environmental Satellite System. This system will take needed operational data and certain long term observations for climate studies. This system will be in place after 2008. The requirements for meteorological variables between 1000 and 1 mb include the following: temperatures accuracys of better than 1.5 K with vertical resolutions of 1 to 5 km from the ground to the mesosphere. The local revisit time is 6 hours and horizontal cell size of about 50 km. The NPOESS also has operational requirements for total column and profile ozone, aerosols and winds

Chemical Measurement Requirements

Atmospheric composition measurements form a cornerstone of any global change strategy. Chemical and dynamical measurements must be made in both the stratosphere and the troposphere. Indeed, chemical measurements around the upper troposphere and lower stratosphere should be among those with the highest priority.

Science questions

The science questions for stratospheric processes are mostly focused on the changes in the stratosphere expected to take place as anthropogenic pollutants accumulate in the middle atmosphere. Greenhouse gases are expected to substantially increase during the EOS period. Stratospheric halogens are expected to increase until 1999, then level off and slowly decline as a result of international regulations. The increases in these gases should produce chemical and dynamical changes. The magnitude of the stratospheric cooling in response to increasing greenhouse gases should far exceed the tropospheric warming because there are fewer feedback mechanisms (such as clouds) which buffer the radiative interaction. Increases in chlorine and bromine will cause decreases in stratospheric ozone. The ozone decrease could be exacerbated by colder lower stratospheric temperatures caused by increasing greenhouse gas concentrations. For example, the colder stratospheric temperatures may lead to an expansion of the extent of PSCs and, hence, polar ozone depletion.

The complex chemistry of the stratosphere can only be understood in detail by measuring a broad range of species over varying conditions with global coverage and over at least an annual cycle. The first area which merits further observational and theoretical study is polar chemistry processes. Direct, simultaneous measurements of HOCl (or a proxy such as ClO), HNO_3, and N_2O_5 are critical since these gases are believed to be involved in PSC surface chemistry. Also, polar night observations, above 20 km, of the chemically-active species, along with PSC measurements, are needed in understanding polar ozone depletion. These regions are not presently accessible with balloons and aircraft.

In order to understand the large ozone depletion at midlatitudes, simultaneous measurements of N_2O_5, HOCl, HNO_3, and HCl are needed to assess the role of heterogeneous chemistry on background aerosols. Since OH and HO_2 drive the chemistry of the lower stratosphere, global measurements of these gases are required to evaluate ozone losses, especially any zonal asymmetries. Also, lower mesosphere observations of OH and HO_2, along with O_3 and temperature, are likely to be key links in understanding the large O_3 decrease expected to occur near 40 km as chlorine levels continue to rise. It is clear that full understanding of these changes requires not just O_3 and ClO measurements, but HO_x and NO_x measurements as well. Measurements in the lower mesosphere, where the chemistry is more simple, may provide the best data set for this analysis.

Key trace gas measurements

There are several scientific requirements to address middle-atmosphere chemistry issues:

1) The self-consistency between the source gases and the resulting active reservoir gases needs to be tested for the four major families that are important to ozone chemistry. The four families and most important species measurements required are: oxygen family (O_3), hydrogen family (H_2O, CH_4, OH, HO_2, H_2O_2), nitrogen family (N_2O, NO_2, HNO_3, N_2O_5), and chlorine family ($CFCl_3$, CF_2Cl_2, HCl, ClO, $ClONO_2$). Stratospheric chlorine is predicted by atmospheric models to increase by 20% in the next five years; thus, our understanding of the production and partitioning among the individual family constituents needs to be verified.

2) The changes in the Antarctic/Arctic lower stratosphere constituents during the ozone hole period in the winter and spring need to be monitored. Since significant changes have been detected during the 1980s and 1990s in the polar regions, these geographical areas require special attention and monitoring.

3) There are a few chemical process studies which require investigation as indicated below.

    a. The HO_x family is fundamentally important in stratospheric chemistry, but the database for that group remains one of the poorest in the atmosphere. Global measurements of the latitudinal, seasonal, and diurnal variation in the HO_x family and related species, H_2O and O_3, are needed to address this deficiency.

    b. Models for the past decade have predicted less ozone in the upper stratosphere than is measured. Several species need to be measured in the upper stratosphere to help resolve this difficulty.

c.Models, in general, predict less odd nitrogen in the lower stratosphere than observed. Measurements of odd nitrogen species, NO_2, HNO_3, N_2O_5, and $ClONO_2$ in the lower stratosphere will help to deal with this problem. d. Another odd nitrogen species, HNO_3, is not modeled accurately in the wintertime in the mid-to-high latitudes. A measurement of HNO_3, N_2O_5, H_2O, and aerosols should help confront this problem.

4) Global observations of ozone in the lowermost stratosphere (tropopause to about 20 km) with high horizontal, vertical, and temporal resolution are needed in order to quantify the ozone budget in that region of the atmosphere and, in particular, to determine the spatial and temporal distribution of ozone fluxes from the lowermost stratosphere to the troposphere. These fluxes, which are presently known only to within about a factor of two, are important for the ozone budget of the lowermost stratosphere and are crucial for understanding the ozone budget of the upper troposphere.

Stratospheric Aerosols and PSCs

The remote sensing of the composition of aerosol at midlatitudes is fairly straightforward. However, remote sensing of the composition and phase of the aerosol particles, for the case of the PSCs, is a developing topic of research. One research goal is to see to what extent it is possible to estimate the composition, phase, area, and volume densities from orbital observations. It is known that the volume densities of NAT and ternary particles are different. Carslaw et al. and Drdla et al. have shown that ternary solutions best describe some of the NASA High-Altitude Research Aircraft (ER-2) data. Beginning with a sulfuric acid droplet core, the volume density of the aerosol increases as temperatures become colder. These equilibrium curves were calculated using different amounts of ambient HNO_3 (5, 10, and 15 ppbv). Remote-sensing observations of temperature versus volume density (and/or aerosol extinction) will likely help classify the composition and phase of the PSC particles. It is also known that HNO_3 is incorporated in ternary, nitric acid dihydrate (NAD), and NAT particles as a function of temperature (i.e., curves of temperature versus the equilibrium gas phase of HNO_3 differ for the three compounds). Therefore, the simultaneous observation of aerosol extinction and HNO_3 gas mixing ratios should help one to classify regions of PSCs as to composition and phase.

Since the microphysics of PSC particles is very temperature sensitive, absolute temperatures need to be measured to plus-or-minus 2 K, since curves of temperature versus volume density for NAT, ternary, and NAD particles differ by only a few kelvins. Remote-sensing observations also average over many kilometers along a horizontal ray path. Vertical coverage is usually on the order of several km. Thus, the fine-scale structure of PSCs, as sampled by ER-2 instruments, cannot be resolved by the remote sounder. Another complication is due to present limitations in the theoretical understanding of how PSCs form, which compositions are formed, and the need for additional laboratory work to quantify at cold stratospheric temperatures the rates at which realistic PSC particles convert inactive to active chlorine compounds, and the need for additional laboratory measurements of the refractive indices of PSC and sulfuric droplets. Current research will see to what extent it is possible to refine present capability to quantify the mechanisms of PSC chemistry, as observed from orbit.

Solar Ultraviolet Flux

Solar radiation at wavelengths below about 300 nm is completely absorbed by the Earth's atmosphere and becomes the dominant direct energy input, establishing the composition and temperature through photodissociation, and driving much of the dynamics as well. Even small changes in this ultraviolet irradiance will have important and demonstrable effects on atmospheric ozone. Radiation between roughly 200 and 300 nm is absorbed by ozone and becomes the major loss mechanism for ozone in the middle atmosphere. Likewise, solar radiation < 200 nm is absorbed predominantly by molecular oxygen and becomes a dominant source of ozone in the middle atmosphere, so changes in these ultraviolet wavelengths will have, to first order, an inverse influence on ozone. These two atmospheric processes, driven by solar radiation, become the major natural control for ozone in the Earth's stratosphere and lower thermosphere. To fully understand the ozone distribution will require many coordinated observations and, in particular, a precise measurement of the solar ultraviolet flux.

The visible portion of solar radiation originates in the solar photosphere and has been accurately measured for about fifteen years. Apparently, this radiation varies by only small fractions of one percent over the 11-year activity cycle of the sun, with comparable variation over time scales of a few days. The ultraviolet portion of the solar spectrum comprises only about

1% (approximately 10 W m^{-2}) and originates from higher layers of the photosphere. As we move to shorter and shorter wavelengths, the emission comes from higher and higher layers of the solar atmosphere. Unlike the solar photosphere, these higher levels are much more under the influence of solar activity, as manifested, for example, by increasing magnetic field strength. As the magnetic activity increases or disappears, the solar radiation, especially the ultraviolet, undergoes dramatic variations modulated by the 27-day rotation period of the sun. Near 120 nm the variation over time periods of days to weeks can be as large as 50%, and over the longer 11-year solar cycle the variation can be as large as a factor of two. Toward longer wavelengths, the solar variability decreases to levels of about 10% at 200 nm and finally to only about 1% at 300 nm. Longward of 300 nm, the intrinsic solar variability is probably only on the order 0.1%, roughly commensurate with measurements of total solar radiation.

The challenge during the EOS time period is to provide measurements of the solar ultraviolet with a precision and accuracy capable of tracking the changes in the solar output. Ideally, the instrument will be capable of measuring changes as small as one percent throughout the EOS mission. This requirement is extremely challenging for solar instruments, especially those making observations at the ultraviolet wavelengths, which are notoriously variable. The harsh environment of space, coupled with the energetic solar radiation, rapidly degrades optical surfaces and usually makes the observations suspect. Some manner of in-flight calibration is required to unambiguously separate changes in the instrument response from true solar changes.

Validation of Satellite Measurements

The role of validation of satellite-based chemical measurements cannot be over stressed. Validation measurements, especially measurements of the same species using two different techniques, have proved to be invaluable for understanding satellite trace species measurements. The very successful UARS validation campaign has contributed a great deal to understanding the individual UARS measurements. The validation campaigns perform two major functions. First, they test the ability of a satellite instrument to make a measurement by giving an independent data point to compare against. Second, if the validation measurements are performed as part of a larger, coordinated campaign, the validation measurements done using aircraft and ground-based measurements can be used to link the small-scale geophysical features that they can observe with the large-scale geophysical features observable from space.

Improvements in Meteorological Measurements

Global limb temperature measurements

The tropopause, the boundary between the upper troposphere (UT) and lower stratosphere (LS), is critical for understanding many important processes in the atmosphere. The tropopause is defined by a sharp change in the vertical temperature gradient, taking place over a few hundred meters at most. Below the tropopause, the troposphere is a region of active vertical mixing. Above the tropopause, the stratosphere is very stable with little vertical mixing. The match between these dissimilar regions, troposphere and stratosphere, modulates the processes that permit the exchange of mass, trace gases, momentum, potential vorticity, and energy between the two regions. Unfortunately, present observing systems do not observe the UT-LS region with sufficient detail. The NOAA operational temperature sensors are characterised by vertical resolution of the retrievals of the order of 10-12 km. The detailed structure of the tropopause is much too thin to be seen by operational systems. However, their cross-track scanning capability gives them the ability to observe horizontal scales of about 100 km.

Temperature profiles with much higher vertical resolution can be obtained by observing the atmospheric limb, or horizon. The improvement results from the geometry, since most of the ray path through the atmosphere is within 1-2 km of the lowest, or tangent, point. In addition, the atmospheric signal is seen against the cold background of space. These factors can reduce the height of the vertical weighting functions to 3-4 km, and the effective resolution to ~ 5 km. EOS limb sounders (MLS and the High-Resolution Dynamics Limb Sounder on the EOS Chemistry Mission [CHEM]) will greatly improve the accuracy, precision, and resolution of temperature measurements in the tropopause region. HIRDLS will determine temperatures with a resolution of 1-1.5 km, through a combination of a narrow (1 km) vertical field of view (FOV), low noise, and oversampling. MLS will make limb temperature measurements with a resolution of 2-3 km.

Higher horizontal resolution temperature profiles

Previous limb scanners have retrieved temperatures with higher vertical resolution, but, because the vertical scans are made at a single azimuth relative to the orbital plane, the horizontal resolution was limited to the orbital spacing, spacing, or about 25°. This is sufficient to resolve only about 6 longitudinal waves. However, the UT-LS is a region in which smaller-scale waves from the troposphere are present. This resolution allows all horizontal waves, up to ~ wavenumber 45, to be observed with the appropriate vertical resolution. Furthermore, the high vertical resolution of HIRDLS together with the high horizontal resolution and daily observations will for the first time provide accurate global mapping of the distribution of ozone in the lowermost stratosphere, a measurement that is crucial for determining the distribution of the flux of ozone to the troposphere, which in turn is crucial for understanding the budget of tropospheric ozone.

Recently a technological innovation has been proposed for the MLS instrument. Instead of a single heterodyne receiver, Microwave Monolithic Integrated Circuit (MMIC) arrays have been proposed at two frequencies. The array system (Array MLS [AMLS]) would allow 100-km × 100-km horizontal resolution temperature, ozone, N_2O, and water with lower power and weight. This proposed system is currently being studied by NASA.

References

Brasseur, Guy P.; Orlando, John J.; Tyndall, Geoffrey S. (1999). *Atmospheric Chemistry and Global Change*. Oxford University Press.

Finlayson-Pitts, Barbara J.; Pitts, James N., Jr. (2000). *Chemistry of the Upper and Lower Atmosphere*. Academic Press.

Seinfeld, John H.; Pandis, Spyros N. (2006). *Atmospheric Chemistry and Physics: From Air Pollution to Climate Change* (2nd Ed.). John Wiley and Sons, Inc.

Warneck, Peter (2000). *Chemistry of the Natural Atmosphere* (2nd Ed.). Academic Press.

Wayne, Richard P. (2000). *Chemistry of Atmospheres* (3rd Ed.). Oxford University Press.

3

Threats to Ecosystems and Biodiversity

The term "ecosystem" was coined in 1930 by Roy Clapham, to denote the physical and biological components of an environment considered in relation to each other as a unit. British ecologist Arthur Tansley later refined the term, describing it as "The whole system,... including not only the organism-complex, but also the whole complex of physical factors forming what we call the environment". Tansely later defined the spatial extent of ecosystems using the term "ecotope" in 1939. Central to the ecosystem concept is the idea that living organisms are continually engaged in a set of relationships with every other element constituting the environment in which they exist. Eugene Odum, one of the founders of the science of ecology, stated: "Any unit that includes all of the organisms (ie: the "community") in a given area interacting with the physical environment so that a flow of energy leads to clearly defined trophic structure, biotic diversity, and material cycles (ie: exchange of materials between living and nonliving parts) within the system is an ecosystem."

Ecosystems have become particularly important politically, since the Convention on Biological Diversity (CBD) - ratified by more than 175 countries - defines "the protection of ecosystems, natural habitats and the maintenance of viable populations of species in natural surroundings" as one of the binding commitments of the ratifying countries. This has created the political necessity to spatially identify ecosystems and somehow distinguish among them. The CBD defines an "ecosystem" as a "dynamic complex of plant, animal and micro-organism communities and their non-living environment interacting as a functional unit".

With the need of protecting ecosystems, the political need arose to describe and identify them within a reasonable time and cost-effectively. Vreugdenhil et al. argued that this could be achieved most effectively by using a physiognomic-ecological classification system, as ecosystems are easily recognisable in the field as well as on satellite images. They argued that the structure and seasonality of the associated vegetation, complemented with ecological data (such as elevation, humidity, drainage, salinity of water and characteristics of water bodies), are each determining modifiers that separate partially distinct sets of species. This is true not only for plant species, but also for species of animals, fungi and bacteria. The degree of ecosystem distinction is subject to the physiognomic modifiers that can be identified on an image and/or in the field. Where necessary, specific fauna elements can be added, such as periodic concentrations of animals and the distribution of coral reefs.

Ecosystem Dynamics

Introduction of new elements, whether biotic or abiotic, into an ecosystem tend to have a disruptive effect. In some cases, this can lead to ecological collapse or "trophic cascading" and the death of many species belonging to the ecosystem in question. Under this deterministic vision, the abstract notion of ecological health attempts to measure the robustness and recovery capacity for an ecosystem; i.e. how far the ecosystem is away from its steady state.

Often, however, ecosystems have the ability to rebound from a disruptive agent. The difference between collapse or a gentle rebound is determined by two factors -- the toxicity of the introduced element and the resiliency of the original ecosystem.

Ecosystems are primarily governed by stochastic (chance) events, the reactions they provoke on non-living materials and the responses by organisms to the conditions surrounding them. Thus, an ecosystem results from the sum of myriad individual responses of organisms to stimuli from non-living and living elements in the environment. The presence or absence of populations merely depends on reproductive and dispersal success, and population levels fluctuate in response to stochastic events. As the number of species in an ecosystem is higher, the number of stimuli is also higher. Since the beginning of life, in this vision, organisms have survived continuous change through natural selection of successful feeding, reproductive and dispersal behavior. Through natural selection the planet's species have continuously adapted to change through variation in their biological composition and distribution.

Mathematically it can be demonstrated that greater numbers of different interacting factors tend to dampen fluctuations in each of the individual factors. Given the great diversity among organisms on earth, most of the time, ecosystems only changed very gradually, as some species would disappear while others would move in. Locally, sub-populations continuously go extinct, to be replaced later through dispersal of other sub-populations. Stochastists do recognise that certain intrinsic regulating mechanisms occur in nature. Feedback and response mechanisms at the species level regulate population levels, most notably through territorial behaviour. Andrewatha and Birch suggest that territorial behaviour tends to keep populations at levels where food supply is not a limiting factor. Hence, stochastists see territorial behaviour as a regulatory mechanism at the species level but not at the ecosystem level. Thus, in their vision, ecosystems are not regulated by feedback and response mechanisms from the (eco)system itself and there is no such thing as a balance of nature.

If ecosystems are indeed governed primarily by stochastic processes, they may be somewhat more resilient to sudden change, as each species would respond individually. In the absence of a balance of nature, the species composition of ecosystems would undergo shifts that would depend on the nature of the change, but entire ecological collapse would probably be less frequently occurring events.

The theoretical ecologist Robert Ulanowicz has used information theory tools to describe the structure of ecosystems, emphasising mutual information (correlations) in studied systems. Drawing on this methodology, and prior observations of complex ecosystems, Ulanowicz depicts approaches to determining the stress levels on ecosystems, and predicting system reactions to defined types of alteration in their settings (such as increased or reduced energy flow, and eutrophication.

Ecosystem Ecology

Ecosystem ecology is the integrated study of biotic and abiotic components of ecosystems and their interactions within an ecosystem framework. This science examines how ecosystems work and relates this to their components such as chemicals, bedrock, soil, plants, and animals. Ecosystem ecology examines physical and biological structure and examines how these ecosystem characteristics interact.

The relationship between systems ecology and ecosystem ecology is complex. Much of systems ecology can be considered a subset of ecosystem ecology. Ecosystem ecology also utilises methods that have little to do with the holistic approach of systems ecology. However, systems ecology more actively considers external influences such as economics that usually fall outside the bounds of ecosystem ecology. Whereas ecosystem ecology can be defined as the scientific study of ecosystems, systems ecology is more of a particular approach to the study of ecological systems and phenomena that interact with these systems.

Systems ecology is an interdisciplinary field of ecology, taking a holistic approach to the study of ecological systems, especially ecosystems. Systems ecology can be seen as an application of general systems theory to ecology. Central to the systems ecology approach is the idea that an ecosystem is a complex system exhibiting emergent properties. Systems ecology focuses on interactions and transactions within and between biological and ecological systems, and is especially concerned with the way the functioning of ecosystems can be influenced by human interventions. It uses and extends concepts from thermodynamics and develops other macroscopic descriptions of complex systems.

Habitat

A habitat is an ecological or environmental area that is inhabited by a particular species. It is the natural environment in which an organism lives, or the physical environment that surrounds (influences and is utilised by) a species population. The term "species population" is preferred to "organism" because, while it is possible to describe the habitat of a single black bear, we may not find any particular or individual bear but the grouping of bears that comprise a breeding population and occupy a certain biogeographical area. Further, this habitat could be somewhat different from the habitat of another group or population of black bears living elsewhere. Thus it is neither the species nor the individual for which the term habitat is typically used.

A microhabitat is a physical location that is home to very small creatures, such as woodlice. Microenvironment is the immediate surroundings and other physical factors of an individual plant or animal within its habitat.

Ecological Use

The term "habitat" can be used more easily in ecology. It was originally defined as the physical conditions that surround a species, or species population, or assemblage of species, or community (Clements and Shelford, 1939). Thus, it is not just a species population that has a habitat, but an assemblage of many species living together in the same place that essentially share a habitat. In ecology, the habitat shared by many species is called a biotope. A biome is the set of flora and fauna which live in a habitat and occupy a certain geography.

Habitats can provide greater protection from big animals, for example, a thick undergrowth where an animal such as the Kudu may hide or go unnoticed.

Threats

Habitat destruction is a major factor in causing a species population to decrease, eventually leading to its being endangered, or even to its extinction. Large scale land clearing usually results in the removal of native vegetation and habitat destruction. Bushfires and poor fire management, pest and weed invasion, cyclone and storm damage can also destroy habitat.

One of the roles of national parks, nature reserves and other protected areas is to provide adequate refuge to animals by preserving habitat.

Ecological Niche

In ecology, a niche (pronounced "nich," "neesh" or "nish") is a term describing the relational position of a species or population in its ecosystem. A shorthand definition is that a niche is how an organism makes a living. The ecological niche describes how an organism or population responds to the distribution of resources and competitors (e. g., by growing when resources are abundant, and predators, parasites and pathogens are scarce) and how it in turn alters those same factors (e.g., limiting access to resources by other organisms, acting as a food source for predators and a consumer of prey).

Parameters

The different dimensions, or plot axes, of a niche represent different biotic and abiotic variables. These factors may include descriptions of the organism's life history, habitat, trophic position (place in the food chain), and geographic range. According to the competitive exclusion principle, no two species can occupy the same niche in the same environment for a long time.

The niche concept was popularised by the zoologist G. Evelyn Hutchinson in 1958. Hutchinson wanted to know why there are so many different types of organisms in any one habitat. The full range of environmental conditions (biological and physical) under which an organism can exist describes its fundamental niche. As a result of pressure from, and interactions with, other organisms (e.g. superior competitors), species are usually forced to occupy a niche that is narrower than this, and to which they are mostly highly adapted. This is termed the realised niche. The ecological niche has also been termed by G.E. Hutchinson a "hypervolume." This term defines the multi-dimensional space of resources (i.e., light, nutrients, structure, etc.) available to (and specifically used by) organisms. The term adaptive zone was coined by the paleontologist, George Gaylord Simpson, and refers to a set of ecological niches that may be occupied by a group of species that exploit the same resources in a similar manner.

It should be noted that Hutchinson's "niche" (a description of the ecological space occupied by a species) is subtly different from the "niche" as defined by Grinnell (an ecological role, that may or may not be actually filled by a species-see vacant niches).

Different species can hold similar niches in different locations and the same species may occupy different niches in different locations. The Australian grasslands species, though different from those of the Great Plains grasslands, occupy the same niche. Once a niche is left vacant, other organisms can fill into that position. For example, the niche that was left vacant by the extinction of the tarpan has been filled by other animals (in particular a small horse breed, the konik). When plants and animals are introduced into a new environment, they can occupy the new niches or niches of native organisms, outcompete the indigenous species, and become a serious pest.

Biological Interaction

Biological interactions result from the fact that organisms in an ecosystem interact with each other, in the natural world, no organism is an autonomous entity isolated from its surroundings. It is part of its environment, rich in living and non living elements all of which interact with each other in some fashion. An organism's interactions with its environment are fundamental to the survival of that organism and the functioning of the ecosystem as a whole. Sign-mediated interactions in which molecules serve as signs are the characteristic feature of communicative interactions.

In ecology, biological interactions are the relationships between two species in an ecosystem. These relationships can be categorised into many different classes of interactions based either on the effects or on the mechanism of the interaction. The interactions between two species vary greatly in these aspects as well as in duration and strength. Species may meet once in a generation (e.g. pollination) or live completely within another (e.g. endosymbiosis). Effects may range from one species eating the other (predation), to mutual benefit (mutualism).

The interactions between two species need not be through direct contact. Due to the connected nature of ecosystems, species may affect each other through intermediaries such as shared resources or common enemies.

Neutralism

Neutralism describes the relationship between two species which do interact but do not affect each other. It is to describe interactions where the fitness of one species has absolutely no effect whatsoever on that of other. True neutralism is extremely unlikely and impossible to prove. When dealing with the complex networks of interactions presented by ecosystems, one cannot assert positively that there is absolutely no competition between or benefit to either species.Since true neutralism is rare or nonexistent, its usage is often extended to situations where interactions are merely insignificant or negligible.

Amensalism

Amensalism between two species involves one impeding or restricting the success of the other without being affected positively or negatively by the presence of the other. It is a type of symbiosis. Usually this occurs when one organism exudes a chemical compound as part of its normal metabolism that is detrimental to another organism.

The bread mold Penicillium is a common example of this; penicillium secrete penicillin, a chemical that kills bacteria. A second example is the black walnut tree (Juglans nigra), which secrete juglone, a chemical that harms or kills some species of neighboring plants, from its roots. This interaction may still increase the fitness of the non-harmed organism though, by removing competition and allowing it access to greater scarce resources. In this sense the impeding organism can be said to be negatively affected by the other's very existence, making it a +/- interaction.

Competition

Competition is an interaction between individuals or populations that is mutually detrimental. Synnecrosis is a particular case in which the interaction is so mutually detrimental that it results in death, as in the case of some parasitic relationships. It is a rare and necessarily short-lived condition as evolution selects against it. The term is seldom used.

Antagonism

In antagonistic interactions one species benefits at the expense of another. Predation is an interaction between organisms in which one organism captures biomass from another. It is often used as a synonym for carnivory but in its widest definition includes all forms of one organism eating another, regardless of trophic level (e.g. herbivory), closeness of association (e.g. parasitism and parasitoidism) and harm done to prey (e.g. grazing). Other interactions that cannot be classed as predation however are still possible, such as Batesian mimicry, where an organism bears a superficial similarity of at least one sort, such as a harmless plant coming to mimic a poisonous one.

Ecological Facilitation

The following two interactions can be classed as facilitative. Facilitation describes species interactions that benefit at least one of the participants and cause no harm to either. Facilitations can be categorized as mutualisms, in which both species benefit, or commensalisms, in which one species benefits and the other is unaffected. Much of classic ecological theory (e.g., natural selection, niche separation, metapopulation dynamics) has focused on negative interactions such as predation and competition, but positive interactions (facilitation) are receiving increasing focus in ecological research.

Commensalism

Commensalism benefits one organism and the other organism is neither benefited nor harmed. It occurs when one organism takes benefits by interacting with another organism by which the host orgaism is not affected. A good example is a remora living with a shark. Remoras eat leftover food from the shark. The shark is not affected in the process as remoras eat only leftover food of the shark which doesn't deplete the sharks resources.

Mutualism

Mutualism is an interaction between two or more species, where species derive a mutual benefit, for example an increased carrying capacity. Similar interactions within a species are known as co-operation. Mutualism may be classified in terms of the closeness of association, the closest being symbiosis, which is often confused with mutualism. One or both species involved in the interaction may be obligate, meaning they cannot survive in the short or long term without the other species. Though mutualism has historically received less attention than other interactions such as predation, it is very important subject in ecology. Examples include cleaner fish, pollination and seed dispersal, gut flora and nitrogen fixation by fungi.

Symbiosis

The term symbiosis can be used to describe various degrees of close relationship between organisms of different species. Sometimes it is used only for cases where both organisms benefit, sometimes it is used more generally to describe all varieties of relatively tight relationships, i.e. even parasitism, but not predation. Some even go so far as to use it to describe predation. It can be used to describe relationships where one organism lives on or in another, or it can be used to describe cases where organisms are related by mutual stereotypic behaviors.

In either case symbiosis is much more common in the living world and much more important than is generally assumed. Almost every organism has many internal parasites. A large percentage of herbivores have mutualistic gut fauna that help them digest plant matter, which is more difficult to digest than animal prey. Coral reefs are the result of mutalisms between coral organisms and various types of algae that live inside them. Most land plants and thus, one might say, the very existence of land ecosystems rely on mutualisms between the plants which fix carbon from the air, and Mycorrhyzal fungi which help in extracting minerals from the ground. In fact the evolution of all eukaryotes (plants, animals, fungi, protists) is believed to have resulted from a symbiosis between various sorts of bacteria: endosymbiotic theory.

Food chain

Food chains, also called, food networks and/or trophic networks, describe the feeding relationships between species within an ecosystem. Organisms are connected to the organisms they consume by arrows representing the direction of biomass transfer. It also shows you how the energy from the producer is given to the consumer.Typically a food chain or food web refers to a graph where only connections are recorded, and a food network or ecosystem network refers to a network where the connections are given weights representing the quantity of nutrients or energy being transfered.

Organisms in Food Chains

Primary producers, commonly called the autotrophys, are species capable of producing complex organic substances (essentially "food") from an energy source and inorganic materials. These organisms are typically photosynthetic plants, bacteria or algae, but in rare cases, like those organisms forming the base of deep-sea vent food webs, can be hemitrope icy. Organisms that get their energy by consuming organic substances are called heterotrophys. Heterotrophy include herbivores, which obtain their energy by consuming live plants; carnivores, which obtain energy from eating live animals; as well as detritivores, scavengers and decomposers, which all consume dead biomass. Energy enters the food chain from the sun. Some energy and/or biomass is lost at each stage of the food chain as; feces (solid waste), movement energy and heat energy (especially by birds and mammals). Therefore, only a small amount of energy and biomass is incorporated into consumer's body and transferred to the next feeding level, thus showing a Pyramid of Biomass.

A food chain is the flow of energy from one organism to the next and to the next and to the next. Organisms in a food chain are grouped into tropic levels &mash; from the Greek word for nourishment, tropics &mash; based on how many links they are removed from the primary producers. Tropic levels may consist of either a single species or a group of species that are presumed to share both predators and prey. They usually start with a primary producer and end with a carnivore. Figure 1 is a food chain from a Swedish lake. It can be described as follows: osprey feed on northern pike that feed on perch that eat bleak that feed on freshwater shrimp.

Although they are not shown in this diagram, the base of this food chain is likely phytoplankton. Phytoplankton are autotrophy, and are the base of the food chain by virtue of their ability to photosynthesize. Phytoplankton, as well as attached algae forms the base of most freshwater food chains. It is often the case that biomass of each tropic level decreases from the base of the chain to the top. This is because energy is lost to the environment with each transfer. On average, only 10% of the organism's energy is passed on to its predator. The other 90% is used for the organism's life processes or it is lost as heat to the environment.

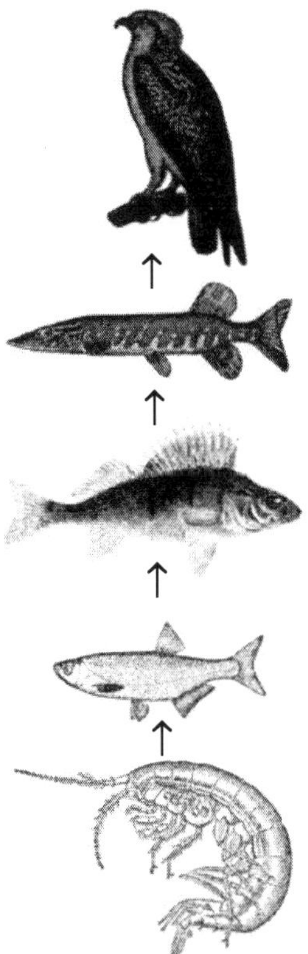

Figure 1. Example of a food chain in a Swedish lake

Graphic representations of the biomass or productivity at each tropic level are called tropic pyramid s. In this food chain for example, the biomass of osprey is smaller than the biomass of pike, which is smaller than the biomass of perch. Some producers, especially phytoplankton, are so productive and have such a high turnover rate that they can actually support a larger biomass of grazers. This is called an inverted pyramid, and can occur when consumers live longer and grow more slowly than the organisms they consume. In this food chain, the productivity of phytoplankton is much greater than that of the zooplankton consuming them. The biomass of the phytoplankton, however, may actually be less than that of the copepods.

Directly linked to this are pyramids of numbers, which show that as the chain is travelled along, the number of consumers at each level drops very significantly, so that a single top consumer (e.g. a Polar Bear) will be supported by literally millions of separate producers (e.g. Phytoplankton). Food chains are overly simplistic as representatives of what typically happens in nature. The food chain shows only one pathway of energy and material transfer. Most consumers feed on multiple species and are, in turn, fed upon by multiple other species. The relations of detritivores and parasites are seldom adequately characterized in such chains as well. The food chain has a producer, consumer, herbivore, carnivore, omnivore, decomposer Arrows in a food web represent an organism getting eaten by another organism.

Food Web

A food web extends the food chain concept from a simple linear pathway to a complex network of interactions. The earliest food webs were published by Victor Summerhayes and Charles Elton in 1923 and Hardy in 1924. Summerhayes and Elton's depicted the interactions of plants, animals and bacteria on Bear Island, Norway, while Hardy's food web showed the interactions of herring and plankton in the North Sea.

The direct steps as shown in the food chain example above seldom reflect reality. This web makes it possible to show much bigger animals (like a seal) eating very small organisms (like plankton). Food sources of most species in an ecosystem are much more diverse, resulting in a complex web of relationships. The grouping of Algae →Protozoa→Oligochaeta→Northern Eider→Arctic Fox is a chain; the whole complex network is a food web.

Biological Adaptation

An adaptation is a characteristic of an organism that has been favored by natural selection and increases the fitness of its possessor. Of course, an adaptation must have been adaptive at some point in an organism's evolutionary history, but such an organism's environment and ecological niche can change over time, leading to adaptations becoming redundant or even a hindrance (maladaptations). Such adaptations are termed vestigial.

"Adaptation" is also sometimes used to refer to a change in an individual organism over the course of its life that makes it more suited to the environment. For an example, see Adaptation (eye). More specifically, however, such changes are referred to as acclimation or acclimatization, the former generally being a very short-term response such as shivering, the latter being a longer-term change such as sun tanning.

There is a great difference between selective adaptation and acclimatization. Adaptation occurs over many generations; it is a gradual process caused by natural selection. Acclimatization generally occurs within a single lifetime and copes with issues that are less threatening. For example, if a human was to move to a higher altitude, respiration and physical exertion would become a problem, but after spending time in high altitude conditions one may acclimate or acclimatize to the pressure and function and no longer notice the change. This ability to acclimate is an adaptation, but not the acclimatization itself.

A counter-adaptation is an adaptation that has evolved due to the selective pressure of another adaptation. This occurs in an evolutionary arms race, where a new adaptation giving one species an advantage is countered by the appearance and spread of a new feature that reduces the effectiveness of the first adaptation. Adaptation is quite a varied subject and their are many main and basic types of adaptation. These include :colour (eg. mating, camoflague),behavioural (eg. finding shade for cooling purposes),functional (eg. shivering),structural (eg. wings)

Theories of Adaptation

The theory of adaptation was first put forth by Jean-Baptiste Lamarck. His theories are also referred to as the inheritance of acquired traits. Lamarck's theory was for a time held as an alternative scientific explanation for evolutionary change observed by Darwin in the The Origin of Species. The classic giraffe analogy offers the best delineation between the two.

- According to Darwin, more long-necked giraffes reproduce than short-necked giraffes and as such giraffes today have long necks.
- According to Lamarck, it was giraffes stretching their necks in response to higher leaves that resulted in giraffes having long necks. (This trait being passed on to the next generation)

Although neither theory in its conception could provide a complete description of the mechanism of transmission of trait variation (i.e., particulate inheritance), many recognized Darwin's theory immediately upon publication as a more complete and empirically supported theory. Modern genetics have since established the fundamental implausibility of Lamarckian inheritance, due to the one-way nature of transcription. However, see epigenetics and Baldwinian evolution for analogous processes in modern evolutionary.

Communities

Ecologists find that within a community many populations are not randomly distributed. This recognition that there was a pattern and process of spatial distribution of species was a major accomplishment of ecology. Two of the most important patterns are open community structure and the relative rarity of species within a community.

Do species within a community have similar geographic range and density peaks? If they do, the community is said to be a closed community, a discrete unit with sharp boundaries known as ecotones. An open community, however, has its populations without ecotones and distributed more or less randomly.

In a forest, where we find an open community structure, there is a gradient of soil moisture. Plants have different tolerances to this gradient and occur at different places along the continuum. Where the physical environment has abrupt transitions, we find sharp boundaries developing between populations. For example, an ecotone develops at a beach separating water and land.

Open structure provides some protection for the community. Lacking boundaries, it is harder for a community to be destroyed in an all or nothing fashion. Species can come and go within communities over time, yet the community as a whole persists. In general, communities are less fragile and more flexible than some earlier concepts would suggest.

Most species in a community are far less abundant than the dominant species that provide a community its name: for example oak-hickory, pine, etc. Populations of just a few species are dominant within a community, no matter what community we examine. Resource partitioning is thought to be the main cause for this distribution.

There are two basic categories of communities: terrestrial (land) and aquatic (water). These two basic types of community contain eight smaller units known as biomes. A biome is a large-scale category containing many communities of a similar nature, whose distribution is largely controlled by climate

Terrestrial Biomes

Tundra and Desert

The tundra a]nd desert biomes occupy the most extreme environments, with little or no moisture and extremes of temperature acting as harsh selective agents on organisms that occupy these areas. These two biomes have the fewest numbers of species due to the stringent environmental conditions. In other words, not everyone can live there due to the specialized adaptations required by the environment.

Tropical Rain Forests

Tropical rain forests occur in regions near the equator. The climate is always warm (between 20° and 25° C) with plenty of rainfall (at least 190 cm/year). The rain forest is probably the richest biome, both in diversity and in total biomass. The tropical rain forest has a complex structure, with many levels of life. More than half of all terrestrial species live in this biome. While diversity is high, dominance by a particular species is low.

While some animals live on the ground, most rain forest animals live in the trees. Many of these animals spend their entire life in the forest canopy. Insects are so abundant in tropical rain forests that the majority have not yet been identified. Charles Darwin noted the number of species found on a single tree, and suggested the richness of the rain forest would stagger the future systematist with the size of the catalogue of animal species found there. Termites are critical in the decomposition and nutrient cycling of wood. Birds tend to be brightly colored, often making them sought after as exotic pets. Amphibians and reptiles are well represented. Lemurs, sloths, and monkeys feed on fruits in tropical rain forest trees. The largest carnivores are the cats (jaguars in South America and leopards in Africa and Asia). Encroachment and destruction of habitat put all these animals and plants at risk.

Epiphytes are plants that grow on other plants. These epiphytes have their own roots to absorb moisture and minerals, and use the other plant more as an aid to grow taller. Some tropical forests in India, Southeast Asia, West Africa, Central and South American are seasonal and have trees that shed leaves in dry season. The warm, moist climate supports high productivity as well as rapid decomposition of detritus.

With its yearlong growing season, tropical forests have a rapid cycling of nutrients. Soils in tropical rain forests tend to have very little organic matter since most of the organic carbon is tied up in the standing biomass of the plants. These tropical soils, termed laterites, make poor agricultural soils after the forest has been cleared.

About 17 million hectares of rain forest are destroyed each year (an area equal in size to Washington state). Estimates indicate the forests will be destroyed (along with a great part of the Earth's diversity) within 100 years. Rainfall and climate patterns could change as a result.

Temperate Forests

The temperate forest biome occurs south of the taiga in eastern North America, eastern Asia, and much of Europe. Rainfall is abundant (30-80 inches/year; 75-150 cm) and there is a well-defined growing season of between 140 and 300 days. The eastern United States and Canada are covered (or rather were once covered) by this biome's natural vegetation, the eastern deciduous forest. Dominant plants include beech, maple, oak; and other deciduous hardwood trees. Trees of a deciduous forest have broad leaves, which they lose in the fall and grow again in the spring.

Sufficient sunlight penetrates the canopy to support a well-developed understory composed of shrubs, a layer of herbaceous plants, and then often a ground cover of mosses and ferns. This stratification beneath the canopy provides a numerous habitats for a variety of insects and birds. The deciduous forest also contains many members of the rodent family, which serve as a food source for bobcats, wolves, and foxes. This area also is a home for deer and black bears. Winters are not as cold as in the taiga, so many amphibian and reptiles are able to survive.

Shrubland (Chaparral)

The shrubland biome is dominated by shrubs with small but thick evergreen leaves that are often coated with a thick, waxy cuticle, and with thick underground stems that survive the dry summers and frequent fires.Shrublands occur in parts of South America, western Australia, central Chile, and around the Mediterranean Sea. Dense shrubland in California, where the summers are hot and very dry, is known as chaparral. This

Mediterranean-type shrubland lacks an understory and ground litter, and is also highly flammable. The seeds of many species require the heat and scarring action of fire to induce germination.

Grasslands

Grasslands occur in temperate and tropical areas with reduced rainfall (10-30 inches per year) or prolonged dry seasons. Grasslands occur in the Americas, Africa, Asia, and Australia. Soils in this region are deep and rich and are excellent for agriculture. Grasslands are almost entirely devoid of trees, and can support large herds of grazing animals. Natural grasslands once covered over 40 percent of the earth's land surface. In temperate areas where rainfall is between 10 and 30 inches a year, grassland is the climax community because it is too wet for desert and too dry for forests.

Most grasslands have now been utilized to grow crops, especially wheat and corn. Grasses are the dominant plants, while grazing and burrowing species are the dominant animals. The extensive root systems of grasses allows them to recover quickly from grazing, flooding, drought, and sometimes fire.

Temperate grasslands include the Russian steppes, the South American pampas, and North American prairies. A tall-grass prairie occurs where moisture is not quite sufficient to support trees. Animal life includes mice, prairie dogs, rabbits, and animals that feed on them (hawks and snakes). Prairies once contained large herds of buffalo and pronghorn antelope, but with human activity these once great herds ahve dwindled.

The savanna is a tropical grassland that contains some trees. The savanna contains the greatest variety and numbers of herbivores (antelopes, zebras, and wildebeests, among others). This environment supports a large population of carnivores (lions, cheetahs, hyenas, and leopards). Any plant litter not consumed by grazers is attacked by termites and other decomposers. Once again, human activities are threatening this biome, reducing the range for herbivores and carnivores.

Deserts

Deserts are characterized by dry conditions (usually less than 10 inches per year; 25 cm) and a wide temperature range. The dry air leads to wide daily temperature fluctuations from freezing at night to over 120 degrees during the day. Most deserts occur at latitudes of 30o N or S where descending air masses are dry. Some deserts occur in the rainshadow of tall mountain ranges or in coastal areas near cold offshore currents. Plants in this biome have developed a series of adaptations (such as succulent stems, and small, spiny, or absent leaves) to conserve water and deal with these temperature extremes. Photosynthetic modifications (CAM) are another strategy to life in the drylands.

The Sahara and a few other deserts have almost no vegetation. Most deserts, however, are home to a variety of plants, all adapted to heat and lack of abundant water (succulents and cacti). Animal life of the Sonoran desert includes arthropods (especially insects and spiders), reptiles (lizards and snakes), running birds (the roadrunner of the American southwest and Warner Brothers cartoon fame), rodents (kangaroo rat and pack rat), and a few larger birds and mammals (hawks, owls, and coyotes).

Taiga (Boreal Forest)

The taiga is a coniferous forest extending across most of the northern area of northern Eurasia and North America. This forest belt also occurs in a few other areas, where it has different names: the montane coniferous forest when near mountain tops; and the temperate rain forest along the Pacific Coast as far south as California. The taiga receives between 10 and 40 inches of rain per year and has a short growing season. Winters are cold and short, while summers tend to be cool. The taiga is noted for its great stands of spruce, fir, hemlock, and pine. These trees have thick protective leaves and bark, as well as needlelike (evergreen) leaves can withstand the weight of accumulated snow. Taiga forests have a limited understory of plants, and a forest floor covered by low-lying mosses and lichens. Conifers, alders, birch and willow are common plants; wolves, grizzly bears, moose, and caribou are common animals. Dominance of a few species is pronounced, but diversity is low when compared to temperate and tropical biomes.

Tundra

The tundra, covers the northernmost regions of North America and Eurasia, about 20% of the Earth's land area. This biome receives about 20 cm (8-10 inches) of rainfall annually. Snow melt makes water plentiful during summer months. Winters are long and dark, followed by very short summers. Water is frozen most of the time, producing frozen soil, permafrost. Vegetation includes no trees, but rather patches of grass and shrubs; grazing musk ox, reindeer, and caribou exist along with wolves, lynx, and rodents. A few animals highly adapted to cold live in the tundra year-round (lemming, ptarmigan). During the summer the tundra hosts numerous insects and migratory animals. The ground is nearly completely covered with sedges and short grasses during the short summer. There are also plenty of patches of lichens and mosses. Dwarf woody shrubs flower and produce seeds quickly during the short growing season. The alpine tundra occurs above the timberline on mountain ranges, and may contain many of the same plants as the arctic tundra.

Climate, Altitude and Terrestrial Biomes

Climate controls biome distribution by an altitudinal gradient and a latitudinal gradient. With increases of either altitude or latitude, cooler and drier conditions occur. Cooler conditions can cause aridity since cooler air can hold less water vapor than can warmer air. Deserts can occur in warm areas due to a blockage of air circulation patterns that form a rain shadow. Warm air rises, producing low pressure areas. Cooler air sinks, producing high pressure areas. The tropics tend to be atmospheric low pressure zones the arctic areas atmospheric highs. Relative humidity is a measure of how much water an air mass at a given temperature can hold. In short, warm air can hold more moisture than can cold air. This basic physical feature of air helps explain the distribution of some of the world's great deserts.

The warm, moist air masses in the tropics rise upward in the atmosphere as they heat. The pressure of air rising forces air in the upper atmosphere to flow away north and south. This air at higher elevations is cooler and loses much of its moisture as rainfall. When the air masses begin to descend they heat up and begin to draw moisture from the lands they descend upon, at 30 degrees north and south of the equator. Many of the world's deserts are at approximately 30 degrees latitude.

Rain shadow deserts also form when cool, dry air masses descend after passing over a tall mountain range, such as the Coast Range and Sierras in California. The Sonoran desert in Arizona is a doubly caused desert, being at 30 degrees latitude as well as in the rain shadow of California mountains. The Tian Shan desert in China is a typical rain shadow desert.

Aquatic Biomes

Conditions in water are generally less harsh than those on land. Aquatic organisms are buoyed by water support, and do not usually have to deal with desiccation. Despite covering 71% of the Earth's surface, areas of the open ocean are a vast aquatic desert containing few nutrients and very little life. Clearcut biome distinctions in water, like those on land, are difficult to make. Dissolved nutrients controls many local aquatic distributions. Aquatic communities are classified into: freshwater (inland) communities and marine (saltwater or oceanic) communities.

Marine Biome

The marine biome contains more dissolved minerals than the freshwater biome. Over 70% of the Earth's surface is covered in water, by far the vast majority of that being saltwater. There are two basic categories to this biome: benthic and pelagic. Benthic communities (bottom dwellers) are subdivided by depth: the shore/shelf and deep sea. Pelagic communities (swimmers or floaters suspended in the water column) include planktonic (floating) and nektonic (swimming) organisms. The upper 200 meters of the water column is the euphotic zone to which light can penetrate.

Coastal Communities

Estuaries are bays where rivers empty into the sea. Erosion brings down nutrients and tides wash in salt water; forms nutrient trap. Estuaries have high production for organisms that can tolerate changing salinity. Estuaries are called "nurseries of the sea" because many young marine fish develop in this protected environment before moving as adults into the wide open seas.

Seashores

Rocky shorelines offer anchorage for sessile organisms. Seaweeds are main photosynthesizers and use holdfasts to anchor. Barnacles glue themselves to stone. Oysters and mussels attach themselves by threads. Limpets and periwinkles either hide in crevices or fasten flat to rocks. Sandy beaches and shores are shifting strata. Permanent residents therefore burrow underground. Worms live permanently in tubes.

Coral Reefs

Areas of biological abundance in shallow, warm tropical waters. Stony corals have calcium carbonate exoskeleton and may include algae. Most form colonies; may associate with zooxanthellae dinoflagellates. Reef is densely populated with animal life. The Great Barrier Reef of Australia suffers from heavy predation by crown-of-thorns sea star, perhaps because humans have harvested its predator, the giant triton.

Oceans

Oceans cover about three-quarters of the Earth's surface. Oceanic organisms are placed in either pelagic (open water) or benthic (ocean floor) categories. Pelagic division is divided into neritic and three levels of pelagic provinces. Neritic province has greater concentration of organisms because sunlight penetrates; nutrients are found here. Epipelagic zone is brightly lit, has much photosynthetic phytoplankton, that support zooplankton that are food for fish, squid, dolphins, and whales. Mesopelagic zone is semi-dark and contains carnivores; adapted organisms tend to be translucent, red colored, or luminescent; for example: shrimps, squids, lantern and hatchet fishes. The bathypelagic zone is completely dark and largest in size; it has strange-looking fish. Benthic division includes organisms on continental shelf (sublittoral), continental slope (bathyal), and the abyssal plain.

Sublittoral zone harbors seaweed that becomes sparse where deeper; most dependent on slow rain of plankton and detritus from sunlit water above. Bathyal zone continues with thinning of sublittoral organisms. Abyssal zone is mainly animals at soil-water interface of dark abyssal plain; in spite of high pressure, darkness and coldness, many invertebrates thrive here among sea urchins and tubeworms.

Thermal vents along oceanic ridges form a very unique community. Molten magma heats seawater to 350oC, reacting with sulfate to form hydrogen sulfide (H2S). Chemosynthetic bacteria obtain energy by oxidizing hydrogen sulfide. The resulting food chain supports a community of tubeworms and clams.

Freshwater Biomes

The freshwater biome is subdivided into two zones: running waters and standing waters. Larger bodies of freshwater are less prone to stratification (where oxygen decreases with depth). The upper layers have abundant oxygen, the lowermost layers are oxygen-poor. Mixing between upper and lower layers in a pond or lake occurs during seasonal changes known as spring and fall overturn.

Lakes are larger than ponds, and are stratified in summer and winter. The epilimnion is the upper surface layer. It is warm in summer. The hypolimnion is the cold lower layer. A sudden drop in temperature occurs at the middle of the thermocline. Layering prevents mixing between the lower hypolimnion (rich in nutrients) and the upper epilimnion (which has oxygen absorbed from its surface). The epilimnion warms in spring and cools in fall, causing a temporary mixing. As a consequence, phytoplankton become more abundant due to the increased amounts of nutrients.

Life zones also exist in lakes and ponds. The littoral zone is closest to shore. The limnetic zone is the sunlit body of the lake. Below the level of sunlight penetration is the dark profundal zone. At the soil-water interface we find the benthic zone. The term benthos is applied to animals and other organisms that live on or in the benthic zone.

Rapidly flowing, bubbling streams have insects and fish adapted to oxygen-rich water. Slow moving streams have aquatic life more similar to lake and pond life.

Community Density and Stability

Communities are made up of species adapted to the conditions of that community. Diversity and stability help define a community and are important in environmental studies. Species diversity decreases as we move away from the tropics. Species diversity is a measure of the different types of organisms in a community (also referred to as species richness). Latitudinal diversity gradient refers to species richness decreasing steadily going away from the equator. A hectare of tropical rain forest contains 40-100 tree species, while a hectare of temperate zone forest contains 10-30 tree species. In marked contrast, a hectare of taiga contains only a paltry 1-5 species! Habitat destruction in tropical countries will cause many more extinctions per hectare than it would in higher latitudes.

Environmental stability is greater in tropical areas, where a relatively stable/constant environment allows more different kinds of species to thrive. Equatorial communities are older because they have been less disturbed by glaciers and other climate changes, allowing time for new species to evolve. Equatorial areas also have a longer growing season.

The depth diversity gradient is found in aquatic communities. Increasing species richness with increasing water depth. This gradient is established by environmental stability and the increasing availability of nutrients.

Community stability refers to the ability of communities to remain unchanged over time. During the 1950s and 1960s, stability was equated to diversity: diverse communities were also stable communities. Mathematical modeling during the 1970s showed that increased diversity can actually increase interdependence among species and lead to a cascade effect when a keystone species is removed. Thus, the relation is more complex than previously thought.

Change in Communities

Biological communities, like the organisms that comprise them, can and do change over time. Ecological time focuses on community events that occur over decades or centuries. Geological time focuses on events lasting thousands of years or more.

Community succession is the sequential replacement of species by immigration of new species and local extinction of older ones following a disturbance that creates unoccupied habitats for colonization. The initial rapid colonizer species are the pioneer community. Eventually a climax community of more or less stable but slower growing species eventually develops.

During succession productivity declines and diversity increases. These trends tend to increase the biomass (total weight of living tissue) in a community. Succession occurs because each community stage prepares the environment for the stage following it.

Primary succession begins with bare rock and takes a very long time to occur. Weathering by wind and rain plus the actions of pioneer species such as lichens and mosses begin the buildup of soil. Herbaceous plants, including the grasses, grow on deeper soil and shade out shorter pioneer species. Pine trees or deciduous trees eventually take root and in most biomes will form a climax community of plants that are stabile in the environment. The young produced by climax species can live in that environment, unlike the young produced by successional species.

Secondary succession occurs when an environment has been disturbed, such as by fire, geological activity, or human intervention (farming or deforestation in most cases). This form of succession often begins in an abandoned field with soil layers already in place. Compared to primary succession, which must take long periods of time to build or accumulate soil, secondary succession occurs rapidly. The herbaceous pioneering plants give way to pines, which in turn may give way to a hardwood deciduous forest (in the classical old field succession models developed in the eastern deciduous forest biome).

Early researchers assumed climax communities were determined for each environment. Today we recognize the outcome of competition among whatever species are present as establishing the climax community.

Climax communities tend to be more stable than successional communities. Early stages of succession show the most growth and are most productive. Pioneer communities lack diversity, make poor use of inputs, and lose heat and nutrients. As succession proceeds, species variety increases and nutrients are recycled more. Climax communities make fuller use of inputs and maintain themselves, thus, they are more stable. Human activity (such as clearing a climax forest community to establish a farm field consisting of a cultivated pioneering species, say corn or wheat) replaces climax communities with simpler communities.

Communities are composed of species that evolve, so the community must also evolve. Comparing marine communities of 500 million years ago with modern communities shows modern communities composed of quite different organisms. Modern communities also tend to be more complex, although this may be a reflection of the nature of the fossil record as well as differences between biological and fossil species.

Ecosystem under Threat

The effects on ecosystems could be many and varied as both plants and animals are sensitive to changes in the climate. The effect of climate change may be more dramatic and rapid than any of the species can adapt to in the natural course of events. Species can take hundreds of years to full adapt to changes in climate conditions. As a results the world may lose many species of both plants and animals. However although the changes might be negative for one region, they might prove to be beneficial to other regions.

The following is how each ecosystem could be effected:

— *Forests*: With the effect of climate change many of these species may have to migrate to cooler climates, this mean that the mainly in a northern direction. However as this type of migration normally happens over the course of millennia and not over the period of decades that may be required, the forests could lose species from it composition while some forests types may disappear altogether.

— *Rangelands*: The is the areas over which some species live. As a result of climate change, there could be change to growing seasons and shifts in the boundaries between different habits. This could result in the loss of fauna and flora, as some species will not be able to migrate due to natural or man-made barriers.

— *Polar Ice-caps and Mountain Glaciers*: It has been predicted that with climate change around one third or more of the worlds mountain glaciers could melt. This could effect things like the supply of water to rivers, creating adverse effects for hydroelectric dams and agricultural areas down stream of the glacier. With the melting of the polar ice-caps the sea levels would rise, with an increase in icebergs in the short term. However some areas will experience in decrease in permafrost, which will cause problems for the humans living in those areas effected. Also there would be a decrease in the are covered and by the thickness of sea-ice, which would likely improve the navigability of the Arctic Ocean.

— *Mountain Regions*: Climate change may led some species to move higher up mountains, some species that only survive in certain high regions, could become extinct. In some regions, such as Nepal, could suffer from the decline in food production and fuel resources. Other areas where the emphasis is on recreational use of the area, the rise in either temperature or a fall in rainfall could been economically disastrous for the region.

— *Water Ways and Wetlands*: It has been predicted that climate changes would result in changes to the water temperatures, flow system and levels. This will increase species in high level areas and reduce species in low level areas, some species will thrive and others will become extinct. Another problem that could be faced is a decline in water quality, due to increases in the occurrence and length of floods or droughts.

— *Coastal Areas*: Some effects for these areas could be the increased erosion of shoreline, increased salinity of estuaries and freshwater aquifers and changes to the tidal reaches of rivers. The areas most likely to be effects by this are salt water marches, mangrove areas, coastal wetlands, coral reefs and atolls and river deltas.

— *Oceans*: With the change in atmospheric temperature due to climate change, the way in which the oceans move around the earth will change. It will also impact on the wave climate and the amount of sea ice. All these will have a secondary effect of changing the available nutrients, the productivity of the biology and even the structure and function of the ocean ecosystem. The changes that could happen to the circulation of the ocean could cause very rapid and dramatic changes in climate, adding to the climate changes would already be taking place.

Some areas of the world may experience problems with food production and sustainability. This reasons could be in region no longer having the correct climate for the growth of certain crops, due to drought or to much rain or even an increase in temperature. Although it has been predicted that globally there will not be a substaintial change in the overall food prodcution. It is also thought that by there very nature developed countries will be better equiped to cope with these changes. It would certainly mean that there would be an overall decline in the flexibility in the distribution of crops.

Overpopulation

The century begins, natural resources are under increasing pressure, threatening public health and development. Water shortages, soil exhaustion, loss of forests, air and water pollution, and degradation of coastlines afflict many areas. As the world's population grows, improving living standards without destroying the environment is a global challenge.

Most developed economies currently consume resources much faster than they can regenerate. Most developing countries with rapid population growth face the urgent need to improve living standards. In the past decade in every environmental sector, conditions have either failed to improve, or they are worsening:

Public Health

Unclean water, along with poor sanitation, kills over 12 million people each year, most in developing countries. Air pollution kills nearly 3 million more. Heavy metals and other contaminants also cause widespread health problems. Amount of land lost to farming by degradation equals 2/3 of North America.

Food Supply

Will there be enough food to go around? In 64 of 105 developing countries studied by the UN Food and Agriculture Organisation, the population has been growing faster than food supplies. Population pressures have degraded some 2 billion hectares of arable land—an area the size of Canada and the U.S.

Freshwater

The supply of freshwater is finite, but demand is soaring as population grows and use per capita rises. By 2025, when world population is projected to be 8 billion, 48 countries containing 3 billion people will face shortages.

Coastlines and Oceans

Half of all coastal ecosystems are pressured by high population densities and urban development. A tide of pollution is rising in the world's seas. Ocean fisheries are being overexploited, and fish catches are down.The demand for forest products exceeds sustainable consumption by 25%.

Forests

Nearly half of the world's original forest cover has been lost, and each year another 16 million hectares are cut, bulldozed, or burned. Forests provide over US$400 billion to the world economy annually and are vital to maintaining healthy ecosystems. Yet, current demand for forest products may exceed the limit of sustainable consumption by 25%. 2/3 of the world's species are in decline.

Biodiversity

The earth's biological diversity is crucial to the continued vitality of agriculture and medicine—and perhaps even to life on earth itself. Yet human activities are pushing many thousands of plant and animal species into extinction. Two of every three species is estimated to be in decline.

Global Climate Change

The earth's surface is warming due to greenhouse gas emissions, largely from burning fossil fuels. If the global temperature rises as projected, sea levels would rise by several meters, causing widespread flooding. Global warming also could cause droughts and disrupt agriculture.

Towards a Livable Future

How people preserve or abuse the environment could largely determine whether living standards improve or deteriorate. Growing human numbers, urban expansion, and resource exploitation do not bode well for the future. Without practicing sustainable development, humanity faces a deteriorating environment and may even invite ecological disaster.

Genetic Erosion

Genetic erosion is a process whereby an already limited gene pool of an endangered species of plant or animal diminishes even more when individuals from the surviving population die off without getting a chance to meet and breed with others in their endangered low population.

Genetic erosion occurs because each individual organism has many unique genes which get lost when it dies without getting a chance to breed. Low genetic diversity in a population of wild animals and plants leads to a further diminishing gene pool, inbreeding and a weakening immune system and fast tracks that species towards eventual extinction.

All the world's endangered species are plagued by varying degrees of Genetic Erosion and most need a human assisted breeding program to keep their population viable and to keep them from going extinct in the long run. The more critically endangered the specie is the more magnified the effect of genetic erosion gets when each surviving individual of the species is lost without getting a fair chance to breed.

Genetic erosion gets compounded and accelerated by habitat fragmentation, today most endangered species live in smaller and smaller chunks of fragmented habitat interspersed with human settlements and farmland making it impossible for them to naturally meet and breed with others of their kind, many die off without getting a fair chance to breed and pass on their genes in the living population.

The gene pool of a species or a population is the complete set of unique alleles that would be found by inspecting the genetic material of every living member of that species or population. A large gene pool indicates extensive genetic diversity, which is associated with robust populations that can survive bouts of intense selection. Meanwhile, low genetic diversity can cause reduced biological fitness and an increased chance of extinction.

Processes and Consequences

"A population bottleneck creates a shrinking gene pool that leaves fewer and fewer mating partners. What are the genetic implications? The animals become part of a high stakes poker game -- with a crooked dealer. After beginning with a 52-card deck, the players wind up with, say, five cards that they are dealt over and over. As they begin to inbreed, congenital effects appear, both physical and reproductive. Often abnormal sperm increase;

infertility rises; the birthrate falls. Most perilous in the long run, each animal's immune defense system is weakened. Thus, even if an endangered species in a bottleneck can withstand whatever human development may be eating away at its habitat, it still faces the threat of an epidemic that could well be fatal to the entire population."

Genetic Erosion in Agricultural and Livestock Biodiversity

Genetic erosion in agricultural and livestock biodiversity is the loss of genetic diversity, including the loss of individual genes, and the loss of particular combinants of genes (or gene complexes) such as those manifested in locally adapted landraces of domesticated animals or plants adapted to the natural environment in which they originated. The term genetic erosion is sometimes used in a narrow sense, such as for the loss of alleles or genes, as well as more broadly, referring to the loss of varieties or even species. The major driving forces behind genetic erosion in crops are: variety replacement, land clearing, overexploitation of species, population pressure, environmental degradation, overgrazing, policy and changing agricultural systems.

The main factor, however, is the replacement of local varieties of domestic plants and animals by high yielding or exotic varieties or species. A large number of varieties can also often be dramatically reduced when commercial varieties (including GMOs) are introduced into traditional farming systems. Many researchers believe that the main problem related to agro-ecosystem management is the general tendency towards genetic and ecological uniformity imposed by the development of modern agriculture.

In-situ conservation

With advances in modern science several techniques and safeguards have emerged to check the relentless advance of genetic erosion and the resulting acceleration of endangered species towards extinction. However many of these techniques and safeguards are too expensive yet to be practical, the best way to protect species is to protect their habitat and to let them live in it naturally.

Wildlife sanctuaries and national parks have been created to preserve entire ecosystems with all the web of species which call them home. Wildlife corridors are created to join fragmented habitats to enable endangered species to travel, meet and breed with others of their kind. Scientific conservation and modern wildlife management techniques with the help of scientifically trained staff help manage these protected ecosystems and the wildlife found in them. Wild animals are also translocated and reintroduced to other locations physically when fragmented wildlife habitat is too far and isolated to be able to link it with a wildlife corridor or when local extinction has already occurred.

Ex-situ conservation

Modern policies of the zoo associations and zoos around the world have changed to putting extreme importance on keeping and breeding wild sourced pure species and subspecies of animals and birds in their registered endangered species breeding programs which will have a chance to be reintroduced and survive in the wild. Main objectives of zoos today has changed to breed pure breed species and subspecies to assist conservation efforts in the wild. Zoos do this by maintaining extremely detailed scientific breeding records i.e. studbooks and loaning their pure breed wild animals and birds to other zoos around the country and indeed globally for breeding to safeguard against inbreeding and hybrids which are considered genetically compromised thus not fit for reintroduction in the wild and in the case of unnaturally found hybrids also to guard against genetic pollution in naturally evolved, region specific, pure wild stocks.

Costly and sometimes controversial ultra modern Ex-situ conservation techniques have emerged for saving the genetic biodiversity on our planet and the diversity in their gene pool by guarding against Genetic erosion through modern concepts like seedbanks, sperm banks, tissue banks, etc. Genetic diversity, DNA, Sperms, eggs, embryos can now be frozen and kept in special banks and laboratories which are sometimes called Modern Noha's Ark or Frozen Zoos where modern cryopreservation techniques are used to freeze these living materials and yet keep them alive by storing them submerged in liquid nitrogen tanks. Thus preserved material can then be used for Artificial insemination, in vitro fertilisation, embryo transfer and cloning etc. to protect diversity in the gene pool of critically endangered species.

It is today possible to save Endangered species from Extinction by preserving parts like tissue, sperms, eggs etc. even after the death of a critically endangered animal or collected from one found freshly dead in captivity or from wild and resurrect it with the help of cloning and give it another chance to breed its genes into the living population of the respective species which is threatened with extinction. Resurrection of dead critically endangered wildlife with the help of cloning is still being perfected and is still too expensive to be practical but with time and advancement is science it may well become a routine procedure in the near future. However Modern Noha's Ark or Frozen Zoos and the use of modern cryopreservation techniques makes lot of sense to preserve living material cheaply which future generations of mankind may well use to diversify limited gene pools of endangered species.

Habitat destruction

Habitat destruction is the process in which natural habitat is rendered functionally unable to support the species originally present. In this process, plants and animals which previously used the site are displaced or destroyed, reducing biodiversity. Agriculture is the principal cause of habitat destruction. Other important causes of habitat destruction include mining, logging, trawling and urban sprawl. Habitat destruction is currently ranked as the most important cause of species extinction worldwide.

It is a process of environmental change important in evolution and conservation biology. As the name implies, it describes the emergence of discontinuities (fragmentation) in an organism's preferred environment (habitat). Habitat fragmentation can be caused by geological processes that slowly alter the layout of the physical environment or by human activity such as land conversion, which alters the environment on a rapid time scale. The former is suspected of being one of the major causes of speciation. The latter is causative in extinctions of many species.

This term, or the terms "loss of habitat" and "habitat reduction", can also be used in a wider sense including loss of habitat due to other factors, such as water and noise pollution. Habitat destruction is human-induced habitat change that results in a reduction of natural habitat. This includes conversion of land to agriculture, urban sprawl, infrastructure development, and other anthropogenic changes to the characteristics of land. Habitat degradation, fragmentation, and pollution are subsets of the broader category of habitat destruction; these do not necessarily involve overt destruction of habitat, yet they cause many of the same results.

Effects

In the simplest terms, when a habitat is destroyed, the plants, animals, and other organisms that occupied the habitat have a reduced carrying capacity so that populations decline and extinction becomes more likely. The single greatest threat to species worldwide is the loss of habitat. Temple found that 82% of endangered bird species were significantly threatened by habitat loss. Habitat destruction, often sugar-coated by the phrase of "land-use change", is the primary cause of loss of biodiversity.

Geography

Biodiversity hotspots are mostly tropical regions that feature high concentrations of endemic species and, when all hotspots are combined, may contain over half of the world's terrestrial species. These hotspots are suffering enormous habitat loss, as each hotspot has lost at least 70% of its primary vegetation.

Most of the natural habitat on islands and in areas of high human population density has already been destroyed. Islands suffering extreme habitat destruction include New Zealand, Madagascar, the Philippines, and Japan. South and east Asia-especially China, India, Malaysia, Indonesia, and Japan-and many areas in west Africa have extremely dense human populations that allow little room for natural habitat. Marine areas close to highly populated coastal cities also face degradation of their coral reefs or other marine habitat. These areas include the eastern coasts of Asia and Africa, northern coasts of South America, and the Caribbean Sea and its associated islands.

Regions of unsustainable agriculture and/or unstable governments, which may go hand-in-hand, typically experience high rates of habitat destruction. Central America, Sub-Saharan Africa, and the Amazonian tropical rainforest areas of South America are the main regions with unsustainable agricultural practices or government mismanagement.

Areas of high agricultural output tend to have the highest extent of habitat destruction. In the U.S., less than 25% of native vegetation remains in many parts of the East and Midwest. Only 15% of land area remains unmodified by human activities in all of Europe.

Ecosystems

Tropical rainforests have received most of the attention concerning the destruction of habitat, and for good reason: From the approximately 16 million square kilometers of tropical rainforest habitat that originally existed worldwide, less than 9 million square kilometers remain today. The current rate of deforestation is 160,000 square kilometers per year, which equates to a loss of approximately 1% of original forest habitat each year.

Other forest ecosystems have suffered as much or more destruction as tropical rainforests. Farming and logging have severely disturbed at least 94% of temperate broadleaf forests; many old growth forest stands have lost more than 98% of their previous area due to human activities. Tropical deciduous dry forests are easier to clear and burn and are more suitable for agriculture and cattle ranching than tropical rainforests; consequently, less than 0.1% of dry forests in Central America's Pacific Coast and less than 8% in Madagascar remain from their original extents.

Plains and desert areas have been degraded to a lesser extent. Only 10-20% of the world's drylands, which include temperate grasslands, savannas, and shrublands, scrub and deciduous forests, have been somewhat degraded. But included in that 10-20% of land is the approximately 9 million square kilometers of seasonally dry-lands that humans have converted to deserts through the process of desertification. The tallgrass prairies of North America, on the other hand, have less than 3% of natural habitat remaining that has not been converted to farmland.

Wetlands and marine areas have endured high levels of habitat destruction. More than 50% of wetlands in the U.S. have been destroyed in just the last 200 years. Between 60% and 70% of Europcan wetlands have been completely destroyed. About one-fifth (20%) of marine coastal areas have been highly modified by humans. One-fifth of coral reefs have also been destroyed, and another fifth has been severely degraded by overfishing, pollution, and invasive species; 90% of the Philippines' coral reefs alone have been destroyed. Finally, over 35% mangrove ecosystems worldwide have been destroyed.

Impact on Human Population

Habitat destruction vastly increases an area's vulnerability to natural disasters like flood and drought, crop failure, spread of disease, and water contamination. On the other hand, a healthy ecosystem with good management practices will reduce the chance of these events from happening, or will at least mitigate adverse impacts.

Agricultural land can actually suffer from the destruction of the surrounding landscape. Over the past 50 years, the destruction of habitat surrounding agricultural land has degraded approximately 40% of agricultural land worldwide via erosion, salinisation, compaction, nutrient depletion, pollution, and urbanisation. Humans also lose direct uses of natural habitat when habitat is destroyed. Aesthetic uses such as birdwatching, recreational uses like hunting and fishing, and ecotourism usually rely upon virtually undisturbed habitat. Many people value the complexity of the natural world and are disturbed by the loss of natural habitats and animal or plant species worldwide.

Probably the most profound impact that habitat destruction has on people is the loss of many valuable ecosystem services. Habitat destruction has altered nitrogen, phosphorus, sulfur, and carbon cycles, which has increased the frequency and severity of acid rain, algal blooms, and fish kills in rivers and oceans and contributed tremendously to global climate change. One ecosystem service whose significance is becoming more realised is climate regulation. On a local scale, trees provide windbreaks and shade; on a regional scale, plant transpiration recycles rainwater and maintains constant annual rainfall; on a global scale, plants (especially trees from tropical rainforests) from around the world counter the accumulation of greenhouse gases in the atmosphere by sequestering carbon dioxide through photosynthesis. Other ecosystem services that are diminished or lost altogether as a result of habitat destruction include watershed management, nitrogen fixation, oxygen production, pollination, waste treatment (i.e., the breaking down and immobilisation of toxic pollutants), and nutrient recycling of sewage or agricultural runoff.

The loss of trees from the tropical rainforests alone represents a substantial diminishing of the earth's ability to produce oxygen and use up carbon dioxide. These services are becoming even more important as increasing carbon dioxide levels is one of the main contributors to global climate change.

The loss of biodiversity may not directly affect humans, but the indirect effects of losing many species as well as the diversity of ecosystems in general are enormous. When biodiversity is lost, the environment loses many species that provide valuable and unique roles to the ecosystem. The environment and all its inhabitants rely on biodiversity to recover from extreme environmental conditions. When too much biodiversity is lost, a catastrophic event such as an earthquake, flood, or volcanic eruption could cause an ecosystem to crash, and humans would obviously suffer from that. Loss of biodiversity also means that humans are losing animals that could have served as biological control agents and plants that could potentially provide higher-yielding crop varieties, pharmaceutical drugs to cure existing or future diseases or cancer, and new resistant crop varieties for agricultural species susceptible to pesticide-resistant insects or virulent strains of fungi, viruses, and bacteria.

The negative effects of habitat destruction usually impact rural populations more directly than urban populations. Across the globe, poor people suffer the most when natural habitat is destroyed, because less natural habitat means less natural resources per capita, yet wealthier people and countries simply have to pay more to continue to receive more than their per capita share of natural resources.

Another way to view the negative effects of habitat destruction is to look at the opportunity cost of keeping an area undisturbed. In other words, what are people losing out on by taking away a given habitat? A country may increase its food supply by converting forest land to row-crop agriculture, but the value of the same land may be much larger when it can supply natural resources or services such as clean water, timber, ecotourism, or flood regulation and drought control.

Solutions

In most cases of tropical deforestation, three to four underlying causes are driving two to three proximate causes. This means that a universal policy for controlling tropical deforestation would not be able to address the unique combination of proximate and underlying causes of deforestation in each country. Before any local, national, or international deforestation policies are written and enforced, governmental leaders must acquire a detailed understanding of the complex combination of proximate causes and underlying driving forces of deforestation in a given area or country. This concept, along with many other results about tropical deforestation from the Geist and Lambin study, can easily be applied habitat destruction in general.

Governmental leaders need to take action by addressing the underlying driving forces, rather than merely regulating the proximate causes In a broader sense, governmental bodies at a local, national, and international scale need to emphasise the following:

1. Considering the many irreplaceable ecosystem services provided by natural habitats,
2. Protecting remaining intact sections of natural habitat,
3. Educating the public about the importance of natural habitat and biodiversity,

4. Developing family planning programs in areas of rapid population growth,
5. Finding ways to increase agricultural output than simply increasing the total land in production,
6. Preserving habitat corridors to minimise prior damage from fragmented habitats.

Water and Environmental Issues

Along with access issues comes use issues. The Coca Cola example noted above highlighted one issue of luxury versus needs. Another issue is the efficient (or inefficient) use of water in industrial agriculture, factories and plants.

It takes a great deal of water to manufacture our goods:

— 1 newspaper takes 150 gallons

— 1 liter of orange juice takes 1000 gallons

— 1 pound of beef takes 2500 gallons

— 1 new car takes 40,000 gallons

Food First, mentioned above, charges that "While transnational corporations over-exploit water resources as they expand industrial and agricultural capacity, they pollute the water table through pollution or overuse. Meanwhile developing countries-under onerous lending requirements enforced by the World Bank-have had to aggressively export their way out of debt, devastating watersheds and placing water supplies in danger."

Quoting them further, and at length:

> In the race to compete for foreign direct investment, countries are stripping their environmental laws and protection of natural resources, including water protection. In some cases, such as the world's 850 free trade zones, they either look the other way as environmental laws are broken and waters are criminally polluted or actually set lower standards in these zones than for the rest of the country.

Throughout Latin America and Asia, massive industrialisation in rural communities is affecting the balance between humans and nature. Water use is being diverted from agriculture to industry. Huge corporate factories are moving up the rivers of the Third World, sucking them dry as they go.

Agribusinesses growing crops for export are claiming more of the water once used by family and peasant farmers for food self-sufficiency. The global expansion in mining and manufacturing is increasing the threat of pollution of underground water supplies and contaminating the aquifers that provide more than 50 percent of domestic supplies in most Asian countries.

To feed the voracious global consumer market, China has transformed its entire economy, massively diverting water use from communities and local farming to its burgeoning industrial sector. As the big industrial wells consume more water, millions of Chinese farmers have found their local wells pumped dry. Eighty percent of China's major rivers are now so degraded, they no longer support fish. Economic globalisation and the policies that drive it are proving to be totally unsustainable.

Climate Change and Water Security

Climate change is going to increase water security. Many of the world's most water-stressed areas will get less water, and water flows will become less predictable and more subject to extreme events. Among the projected outcomes:

— Marked reductions in water availability in East Africa, the Sahel and Southern Africa as rainfall declines and temperature rises, with large productivity losses in basic food staples. Projections for rainfed areas in East Africa point to potential productivity losses of up to 33% in maize and more than 20% for sorghum and 18% for millet.

— The disruption of food production systems exposing an additional 75-125 million people to the threat of hunger.

— Accelerated glacial melt, leading to medium term reductions in water availability across a large group of countries in East Asia, Latin America and South Asia.

— Disruptions to monsoon patterns in South Asia, with the potential for more rain but also fewer rainy days and more people affected by drought.

— Rising sea levels resulting in freshwater losses in river delta systems in countries such as Bangladesh, Egypt and Thailand.

For a number of years now, we have heard of predictions that future wars will be fought over control of essential resources, such as water. To some extent, most wars have already been about that. However, in terms of water itself, some experts question this prediction. Inter Press Service (IPS) notes a number of experts disagree with the view that future wars will be over water, and instead feel it is mismanagement of water resources which is the issue, not scarcity (which is the underlying assumption for the prediction of such wars.)

That same IPS article quotes Arunabha Ghosh, co-author of the United Nations Human Development Report 2006 themed on water management who says, "Water wars make good newspaper headlines but cooperation (agreements) don't.... there are plenty of bilateral, multilateral and trans-boundary agreements for water-sharing-all or most of which do not make good newspaper copy."

Others have noted that there are many more examples of cooperation than conflict in regions with shard water interests. The Stockholm International Water Institute opines that "10- to 20-year-old arguments about conflict over water are still being recycled."

At the same time there have been various incidents that fuel the fear of water-related wars, such as Israel's recent bombing of the Lebanese water pipelines from the Litani River to farmland along the coastal plain and parts of the Bekaa Valley, and the conflict in Sri Lanka where the rebel group diverted a canal.

International Agreements and Action

Access to fresh water is becoming a political problem, rather than a technical one, with lots of questions on the best way for countries to provide it. The Millennium Development Goals, a number of targets to help alleviate poverty around the world by 2015, includes the aim to "reduce by half the proportion of people without sustainable access to safe drinking water." A number of international meetings have taken place in recent years.

For example, March 17-22, 2000, saw the Second World Water Forum, which tried to address many issues. The types of topics addressed included the following:

— Water as a human right
— Water Management-not water scarcity-as the problem
— Call for a new Water Ethic; That water is a management problem, a cultural problem, rather than a resource problem in most cases
— Governments should participate in people's projects rather than people participating in governments' projects
— Water culture-and gender. Female involvement will be important. Women are often more sensitive to cultural and other issues which will be important.

— Privatisation-water should maintain a common property resource, common heritage of all. However, there may be costs associated with being able to provide the infrastructure and services in a sustainable way.

— Eco-sanitation: Turning waste into a resource

— Rainwater Harvesting

Some activists were concerned about the corporate agenda in water privatisation. However, as per the final declaration of the water forum, water security was defined to mean that "freshwater, coastal and related ecosystems are protected and improved; that sustainable development and political stability are promoted, that every person has access to enough safe water at an affordable cost to lead a healthy and productive life and that the vulnerable are protected from the risks of water-related hazards."

Threats To Global Biodiversity

Biological diversity refers to the variety and variability among living organisms and the ecological complexes in which they occur. Diversity can be defined as the number of different items and their relative frequency. For biological diversity, these items are organised at many levels, ranging from complete ecosystems to the chemical structures that are the molecular basis of heredity. Thus, the term encompasses different ecosystems, species, genes, and their relative abundance. Or to paraphrase: number and variety of species, ecological systems, and the genetic variability they contain. The simplest representation is:

Group	Number of Described Species
Bacteria and blue-green algae	4,760
Fungi	46,983
Algae	26,900
Bryophytes (mosses and liverworts)	17,000
Gymnosperms (conifers)	750
Angiosperms (flowering plants)	250,000
Protozoans	30,800
Sponges	5,000
Corals and Jellyfish	9,000
Roundworms and earthworms	24,000
Crustaceans	38,000
Insects	751,000
Other Arthropods and minor invertebrates	132,461

Mollusks	50,000
Starfish	6,100
Fishes (teleosts)	19,056
Amphibians	4,184
Reptiles	6,300
Birds	9,198
Mammals	4,170
Total	***1,435,662***

Since 1986 the terms and the concept have achieved widespread use among biologists, environmentalists, political leaders, and concerned citizens worldwide. It is generally used to equate to a concern for the natural environment and nature conservation. This use has coincided with the expansion of concern over extinction observed in the last decades of the 20th century.

The term "natural heritage" pre-dates "biodiversity", though it is a less scientific term and more easily comprehended in some ways by the wider audience interested in conservation. "Natural Heritage" was used when Jimmy Carter set up the Georgia Heritage Trust while he was governor of Georgia; Carter's trust dealt with both natural and cultural heritage. It would appear that Carter picked the term up from Lyndon Johnson, who used it in a 1966 Message to Congress. "Natural Heritage" was picked up by the Science Division of The Nature Conservancy when, under Jenkins, it launched in 1974 the network of State Natural Heritage Programs. When this network was extended outside the USA, the term "Conservation Data Center" was suggested by Guillermo Mann and came to be preferred.

The most straightforward definition is "variation of life at all levels of biological organisation". A second definition holds that biodiversity is a measure of the relative diversity among organisms present in different ecosystems. "Diversity" in this definition includes diversity within a species and among species, and comparative diversity among ecosystems.

A third definition that is often used by ecologists is the "totality of genes, species, and ecosystems of a region". An advantage of this definition is that it seems to describe most circumstances and present a unified view of the traditional three levels at which biodiversity has been identified:

— genetic diversity—diversity of genes within a species. There is a genetic variability among the populations and the individuals of the same species.

— species diversity—diversity among species in an ecosystem. "Biodiversity hotspots" are excellent examples of species diversity.

— ecosystem diversity—diversity at a higher level of organisation, the ecosystem. Diversity of habitat in a given unit area. To do with the variety of ecosystems on Earth.

The 1992 United Nations Earth Summit in Rio de Janeiro defined "biodiversity" as "the variability among living organisms from all sources, including, 'inter alia', terrestrial, marine, and other aquatic ecosystems, and the ecological complexes of which they are part: this includes diversity within species, between species and of ecosystems". This is, in fact, the closest thing to a single legally accepted definition of biodiversity, since it is the definition adopted by the United Nations Convention on Biological Diversity.

If the gene is the fundamental unit of natural selection, according to E. O. Wilson, the real biodiversity is genetic diversity. For geneticists, biodiversity is the diversity of genes and organisms. They study processes such as mutations, gene exchanges, and genome dynamics that occur at the DNA level and generate evolution.

For ecologists, biodiversity is also the diversity of durable interactions among species. It not only applies to species, but also to their immediate environment (biotope) and their larger ecoregion. In each ecosystem, living organisms are part of a whole, interacting with not only other organisms, but also with the air, water, and soil that surround them..

Biodiversity is a broad concept, so a variety of objective measures have been created in order to empirically measure biodiversity. Each measure of biodiversity relates to a particular use of the data.

For practical conservationists, this measure should quantify a value that is broadly shared among locally affected people. For others, a more economically defensible definition should allow the ensuring of continued possibilities for both adaptation and future use by people, assuring environmental sustainability.

As a consequence, biologists argue that this measure is likely to be associated with the variety of genes. Since it cannot always be said which genes are more likely to prove beneficial, the best choice for conservation is to assure the persistence of as many genes as possible. For ecologists, this latter approach is sometimes considered too restrictive, as it prohibits ecological succession.

Biodiversity Distribution

Biodiversity is not distributed evenly on Earth. It is consistently richer in the tropics and in other localised regions such as the California Floristic Province. As one approaches polar regions one generally finds fewer species. Flora and fauna diversity depends on climate, altitude, soils and the presence of other species. In the year 2006 large numbers of the Earth's species were formally classified as rare or endangered or threatened species; moreover, many scientists have estimated that there are millions more species actually endangered which have not yet been formally recognised. About 40 percent of the 40,177 species assessed using the IUCN Red List criteria, are now listed as threatened species with extinction—a total of 16,119 species.

A biodiversity hotspot is a region with a high level of endemic species. These biodiversity hotspots were first identified by Dr. Norman Myers in two articles in the scientific journal *The Environmentalist*. Dense human habitation tends to occur near hotspots. Most hotspots are located in the tropics and most of them are forests.

Brazil's Atlantic Forest is considered a hotspot of biodiversity and contains roughly 20,000 plant species, 1350 vertebrates, and millions of insects, about half of which occur nowhere else in the world. The island of Madagascar including the unique Madagascar dry deciduous forests and lowland rainforests possess a very high ratio of species endemism and biodiversity, since the island separated from mainland Africa 65 million years ago, most of the species and ecosystems have evolved independently producing unique species different from those in other parts of Africa.

Many regions of high biodiversity (as well as high endemism) arise from very specialised habitats which require unusual adaptation mechanisms. For example the peat bogs of Northern Europe and the alvar regions such as the Stora Alvaret on Oland, Sweden host a large diversity of plants and animals, many of which are not found elsewhere.

During the last century, erosion of biodiversity has been increasingly observed. Some studies show that about one eighth known plant species is threatened with extinction[specify]. Some estimates put the loss at up to 140,000 species per year (based on Species-area theory) and subject to discussion. This figure indicates unsustainable ecological practices, because only a small number of species come into being each year. Almost all scientists acknowledge that the rate of species loss is greater now than at any time in human history, with extinctions occurring at rates hundreds of times higher than background extinction rates.

Destruction of Habitats

Most of the species extinctions from 1000 AD to 2000 AD are due to human activities, in particular destruction of plant and animal habitats. Raised rates of extinction are being driven by human consumption of organic resources, especially related to tropical forest destruction. While most of the species that are becoming extinct are not food species, their biomass is converted into human food when their habitat is transformed into pasture, cropland, and orchards. It is estimated that more than 40% of the Earth's biomass is tied up in only the few species that represent humans, livestock and crops. Because an ecosystem decreases in stability as its species are made extinct, these studies warn that the global ecosystem is destined for collapse if it is further reduced in complexity. Factors contributing to loss of biodiversity are: overpopulation, deforestation, pollution (air pollution, water pollution, soil contamination) and global warming or climate change, driven by human activity. These factors, while all stemming from overpopulation, produce a cumulative impact upon biodiversity.

There are systematic relationships between the area of a habitat and the number of species it can support, with greater sensitivity to reduction in habitat area for species of larger body size and for those living at lower latitudes or in forests or oceans. Some characterise loss of biodiversity not as ecosystem degradation but by conversion to trivial standardised ecosystems (e.g., monoculture following deforestation). In some countries lack of property rights or access regulation to biotic resources necessarily leads to biodiversity loss (degradation costs having to be supported by the community).

Exotic species

The rich diversity of unique species across many parts of the world exist only because they are separated by barriers, particularly large rivers, seas, oceans, mountains and deserts from other species of other land masses, particularly the highly fecund, ultra-competitive, generalist "super-species". These are barriers that could never be crossed by natural processes, except for many millions of years in the future through continental drift. However humans have invented ships and airplanes, and now have the power to bring into contact species that never have met in their evolutionary history, and on a time scale of days, unlike the centuries that historically have accompanied major animal migrations.

The widespread introduction of exotic species by humans is a potent threat to biodiversity. When exotic species are introduced to ecosystems and establish self-sustaining populations, the endemic species in that ecosystem, that have not evolved to cope with the exotic species, may not survive. The exotic organisms may be either predators, parasites, or simply aggressive species that deprive indigenous species of nutrients, water and light. These exotic or invasive species often have features, due to their evolutionary background and new environment, that make them highly competitive; able to become well-established and spread quickly, reducing the effective habitat of endemic species.

As a consequence of the above, if humans continue to combine species from different ecoregions, there is the potential that the world's ecosystems will end up dominated by relatively a few, aggressive, cosmopolitan "super-species".

. *Decline in amphibian populations*

Declines in amphibian populations have been observed since 1980s. Because of the sensitivity of these organisms, they are regarded by many scientists as a marker for the overall health of an ecosystem. Their decline has led to concern about the general current state of biodiversity.

Genetic pollution

Purebred naturally evolved region specific wild species can be threatened with extinction in a big way through the process of genetic pollution i.e. uncontrolled hybridisation, introgression and Genetic swamping which leads to homogenisation or replacement of local genotypes as a result of either a numerical and/or fitness advantage of introduced plant or animal. Nonnative species can bring about a form of extinction of native plants and animals by hybridisation and introgression either through purposeful introduction by humans or through habitat modification, bringing previously isolated species into contact.

These phenomena can be especially detrimental for rare species coming into contact with more abundant ones where the abundant ones can interbreed with them swamping the entire rarer gene pool creating hybrids thus driving the entire original purebred native stock to complete extinction.

Attention has to be focused on the extent of this under appreciated problem that is not always apparent from morphological (outward appearance) observations alone. Some degree of gene flow may be a normal, evolutionarily constructive process, and all constellations of genes and genotypes cannot be preserved however, hybridisation with or without introgression may, nevertheless, threaten a rare species' existence.

Hybridisation and genetics

In agriculture and animal husbandry, green revolution popularised the use of conventional hybridisation to increase yield many folds by creating "high-yielding varieties". Often the handful of breeds of plants and animals hybridised originated in developed countries and were further hybridised with local varieties, in the rest of the developing world, to create high yield strains resistant to local climate and diseases. Local governments and industry since have been pushing hybridisation with such zeal that several of the wild and indigenous breeds evolved locally over thousands of years having high resistance to local extremes in climate and immunity to diseases etc. have already become extinct or are in grave danger of becoming so in the near future. Due to complete disuse because of un-profitability and uncontrolled intentional and unintentional cross-pollination and crossbreeding formerly huge gene pools of various wild and indigenous breeds have collapsed causing widespread genetic erosion and genetic pollution resulting in great loss in genetic diversity and biodiversity as a whole.

A genetically modified organism (GMO) is an organism whose genetic material has been altered using the genetic engineering techniques generally known as recombinant DNA technology. Genetically Modified (GM) crops today have become a common source for genetic pollution, not only of wild varieties but also of other domesticated varieties derived from relatively natural hybridisation. It is being said that genetic erosion coupled with genetic pollution is destroying that needed unique genetic base thereby creating an unforeseen hidden crisis which will result in a severe threat to our food security for the future when diverse genetic material will cease to exist to be able to further improve or hybridise weakening food crops and livestock against more resistant diseases and climatic changes.

Threats to Madagascar's flora and fauna

Madagascar is one of the poorest countries in the world. More than 18 million people live there, and the population is growing rapidly. The vast majority of Malagasy people (as inhabitants of Madagascar are called) eke out an existence supported by slash-and-burn agriculture. Farmers clear a plot of land, cultivate it for a few years until the soil is depleted, then move on to clear another patch of forest. Forest is also burned to provide grazing areas for cattle and logged for construction materials, firewood, and charcoal production.

Humans arrived on the island only about 2,000 years ago but have already left a heavy mark:

— Less than 15 percent of the island is undisturbed.

— Less than 10 percent will be protected by 2008.

— It is estimated that only 10 to 15 percent of the island's habitat remains undisturbed.

— Deforestation has caused massive erosion, as the island's soils wash into the ocean.

— To add to the ecological disaster, many animals are illegally hunted for meat or for the international pet trade.

— On top of it all, global climate change looms as a serious threat.

Michelle Zjhra, of Georgia Southern University, who studies Madagascar's trees, said, "Since I've been collecting, the number of these species has gone way up, which means we're only just finding the tip of the iceberg. We're racing against the clock to document the diversity."

To date about 2.7 percent of Madagascar's land area is officially protected. In response to the ecological crisis, in 2003 the president of Madagascar, Marc Ravalomanana, announced he would triple the amount of land under protection in his country before 2008.

Industrialised nations are the worst polluters

The developed nations are the largest polluters in the world today. They must greatly reduce their over consumption, if we are to reduce pressures on resources and the global environment. The developed nations have the obligation to provide aid and support to developing nations, because only the developed nations have the financial resources and the technical skills for these tasks.

Acting on this recognition is not altruism, but enlightened self-interest: whether industrialised or not, we all have but one lifeboat. No nation can escape from injury when global biological systems are damaged. No nation can escape from conflicts over increasingly scarce resources. In addition, environmental and economic instabilities will cause mass migrations with incalculable consequences for developed and undeveloped nations alike.

Developing nations must realise that environmental damage is one of the gravest threats they face, and that attempts to blunt it will be overwhelmed if their populations go unchecked. The greatest peril is to become trapped in spirals of environmental decline, poverty, and unrest, leading to social, economic and environmental collapse.

Success in this global endeavor will require a great reduction in violence and war. Resources now devoted to the preparation and conduct of war—amounting to over $1 trillion annually—will be badly needed in the new tasks and should be diverted to the new challenges.

A new ethic is required—a new attitude towards discharging our responsibility for caring for ourselves and for the earth. We must recognise the earth's limited capacity to provide for us. We must recognise its fragility. We must no longer allow it to be ravaged. This ethic must motivate a great movement, convince reluctant leaders and reluctant governments and reluctant peoples themselves to effect the needed changes.

References

Grumbine, R. E. (1994). "What is ecosystem management?". *Conservation Biology* 8 (1): 27–38.

Hammond, H. (2009). *Maintaining Whole Systems on the Earth's Crown: Ecosystem-based Conservation Planning for the Boreal Forest.* Slocan Park, BC: Silva Forest Foundation. p. 380.

Whittaker, R. H.; Levin, S. A.; Root, R. B. (1973). "Niche, habitat, and ecotope". *The American Naturalist* 107(955): 321–338.

Holling, C. S. (2004). "Understanding the complexity of economic, ecological, and social systems". *Ecosystems* 4 (5): 390–405.

Wilson, E. O. (1992). *The Diversity of Life.* Harvard University Press. pp. 440.

4

Atmospheric Pollution

The atmosphere of Earth is a layer of gases surrounding the planet Earth that is retained by Earth's gravity. The atmosphere protects life on Earth by absorbing ultraviolet solar radiation, warming the surface through heat retention (greenhouse effect), and reducing temperature extremes between day and night (the diurnal temperature variation).

Atmospheric stratification describes the structure of the atmosphere, dividing it into distinct layers, each with specific characteristics such as temperature or composition. The atmosphere has a mass of about 5×1018 kg, three quarters of which is within about 11 km (6.8 mi; 36,000 ft) of the surface. The atmosphere becomes thinner and thinner with increasing altitude, with no definite boundary between the atmosphere and outer space. An altitude of 120 km (75 mi) is where atmospheric effects become noticeable during atmospheric reentry of spacecraft. The Kármán line, at 100 km (62 mi), also is often regarded as the boundary between atmosphere and outer space. This altitude amounts to 1.57% of the Earth's radius.

Air is the name given to the atmosphere used in breathing and photosynthesis. Dry air contains roughly (by volume) 78.09% nitrogen, 20.95% oxygen, 0.93% argon, 0.039% carbon dioxide, and small amounts of other gases. Air also contains a variable amount of water vapor, on average around 1%. While air content and atmospheric pressure vary at different layers, air suitable for the survival of terrestrial plants and terrestrial animals is currently only known to be found in Earth's troposphere and artificial atmospheres.

Atmospheric Composition

Air is mainly composed of nitrogen, oxygen, and argon, which together constitute the major gases of the atmosphere. The remaining gases are often referred to as trace gases, among which are the greenhouse gases such as water vapor, carbon dioxide, methane, nitrous oxide, and ozone. Filtered air includes trace amounts of many other chemical compounds. Many natural substances may be present in tiny amounts in an unfiltered air sample, including dust, pollen and spores, sea spray, and volcanic ash. Various industrial pollutants also may be present, such as chlorine (elementary or in compounds), fluorine compounds, elemental mercury, and sulfur compounds such as sulfur dioxide [SO_2].

Structure of the Atmosphere

In general, air pressure and density decrease in the atmosphere as height increases. However, temperature has a more complicated profile with altitude, and may remain relatively constant or even increase with altitude in some regions (see the temperature section, below). Because the general pattern of the temperature/altitude profile is constant and recognizable through means such as balloon soundings, the temperature behavior provides a useful metric to distinguish between atmospheric layers. In this way, Earth's atmosphere can be divided (called atmospheric stratification) into five main layers. From highest to lowest, these layers are:

Principal Layers

Exosphere

The outermost layer of Earth's atmosphere extends from the exobase upward. It is mainly composed of hydrogen and helium. The particles are so far apart that they can travel hundreds of kilometers without colliding with one another. Since the particles rarely collide, the atmosphere no longer behaves like a fluid. These free-moving particles follow ballistic trajectories and may migrate into and out of the magnetosphere or the solar wind.

Thermosphere

Temperature increases with height in the thermosphere from the mesopause up to the thermopause, then is constant with height. Unlike in the stratosphere, where the inversion is caused by absorption of radiation by ozone, in the thermosphere the inversion is a result of the extremely low density of molecules. The temperature of this layer can rise to 1,500 °C (2,700 °F), though the gas molecules are so far apart that temperature in the usual sense is not well defined. The air is so rarefied that an individual molecule (of oxygen, for example) travels an average of 1 kilometer between collisions with other molecules. The International Space Station orbits in this layer, between 320 and 380 km (200 and 240 mi). Because of the relative infrequency of molecular collisions, air above the mesopause is poorly mixed compared with air below. While the composition from the troposphere to the mesosphere is fairly constant, above a certain point, air is poorly mixed and becomes compositionally stratified. The point dividing these two regions is known as the turbopause. The region below is the homosphere, and the region above is the heterosphere. The top of the thermosphere is the bottom of the exosphere, called the exobase. Its height varies with solar activity and ranges from about 350–800 km.

Mesosphere

The mesosphere extends from the stratopause to 80–85 km. It is the layer where most meteors burn up upon entering the atmosphere. Temperature decreases with height in the mesosphere. The mesopause, the temperature minimum that marks the top of the mesosphere, is the coldest place on Earth and has an average temperature around -85 °C. At the mesopause, temperatures may drop to -100 °C. Due to the cold temperature of the mesosphere, water vapor is frozen, forming ice clouds. A type of lightning referred to as either sprites or ELVES, form many miles above thunderclouds in the troposphere.

Stratosphere

The stratosphere extends from the tropopause to about 51 km (32 mi; 170,000 ft). Temperature increases with height due to increased absorption of ultraviolet radiation by the ozone layer, which restricts turbulence and mixing. While the temperature may be -60 °C at the tropopause, the top of the stratosphere is much warmer, and may be near freezing. The stratopause, which is the boundary between the stratosphere and mesosphere, typically is at 50 to 55 km. The pressure here is 1/1000 sea level.

Troposphere

The troposphere begins at the surface and extends to between 9 km (30,000 ft) at the poles and 17 km (56,000 ft) at the equator, with some variation due to weather. The troposphere is mostly heated by transfer of energy from the surface, so on average the lowest part of the troposphere is warmest and temperature decreases with altitude. This promotes vertical mixing (hence the origin of its name in the Greek word "t??p?", trope, meaning turn or overturn). The troposphere contains roughly 80% of the mass of the atmosphere. The tropopause is the boundary between the troposphere and stratosphere.

Other Layers

Within the five principal layers determined by temperature are several layers determined by other properties:

The ozone layer is contained within the stratosphere. In this layer ozone concentrations are about 2 to 8 parts per million, which is much higher than in the lower atmosphere but still very small compared to the main components of the atmosphere. It is mainly located in the lower portion of the stratosphere from about 15–35 km (9.3–22 mi; 49,000–110,000 ft), though the thickness varies seasonally and geographically. About 90% of the ozone in our atmosphere is contained in the stratosphere.

The ionosphere, the part of the atmosphere that is ionized by solar radiation, stretches from 50 to 1,000 km (31 to 620 mi; 160,000 to 3,300,000 ft) and typically overlaps both the exosphere and the thermosphere. It forms the inner edge of the magnetosphere. It has practical importance because it influences, for example, radio propagation on the Earth. It is responsible for auroras.

The homosphere and heterosphere are defined by whether the atmospheric gases are well mixed. In the homosphere the chemical composition of the atmosphere does not depend on molecular weight because the gases are mixed by turbulence. The homosphere includes the troposphere, stratosphere, and mesosphere. Above the turbopause at about 100 km (62 mi; 330,000 ft) (essentially corresponding to the mesopause), the composition varies with altitude. This is because the distance that particles can move without colliding with one another is large compared with the size of motions that cause mixing. This allows the gases to stratify by molecular weight, with the heavier ones such as oxygen and nitrogen present only near the bottom of the heterosphere. The upper part of the heterosphere is composed almost completely of hydrogen, the lightest element.

The planetary boundary layer is the part of the troposphere that is nearest the Earth's surface and is directly affected by it, mainly through turbulent diffusion. During the day the planetary boundary layer usually is well-mixed, while at night it becomes stably stratified with weak or intermittent mixing. The depth of the planetary boundary layer ranges from as little as about 100 m on clear, calm nights to 3000 m or more during the afternoon in dry regions.

ATMOSPHERIC PRESSURE

Atmospheric pressure is a direct result of the total weight of the air above the point at which the pressure is measured. This means that air pressure varies with location and time, because the amount (and weight) of air above the earth varies with location and time. However the *average* mass of the air above a square meter of the earth's surface is known to the same high accuracy as the total air mass of 5148.0 teratonnes and area of the earth of 51007.2 megahectares, namely 5148.0/510.072 = 10.093 metric tonnes or 14.356 lbs (mass) per square inch. This is about 2.5% below the officially standardised unit atmosphere (1 atm) of 101.325 kPa or 14.696 psi, and corresponds to the mean pressure not at sea level but at the mean base of the atmosphere as contoured by the earth's terrain.

Were atmospheric density to remain constant with height the atmosphere would terminate abruptly at 7.81 km (25,600 ft). Instead it decreases with height, dropping by 50% at an altitude of about 5.6 km (18,000 ft). For comparison: the highest mountain, Mount Everest, is higher, at 8.8 km, which is why it is so difficult to climb without supplemental oxygen. This pressure drop is approximately exponential, so that pressure decreases by approximately half every 5.6 km (whence about 50% of the total atmospheric mass is within the lowest 5.6 km) and by $1/e = .368$ every 7.64 km, the average scale height of Earth's atmosphere below 70 km. However, because of changes in temperature, average molecular weight, and gravity throughout the atmospheric column, the dependence of atmospheric pressure on altitude is modeled by separate equations for each of the layers listed above.

Even in the exosphere, the atmosphere is still present (as can be seen for example by the effects of atmospheric drag on satellites). The equations of pressure by altitude in the above references can be used directly to estimate atmospheric thickness. However, the following published data are given for reference:

— 50% of the atmosphere by mass is below an altitude of 5.6 km.

— 90% of the atmosphere by mass is below an altitude of 16 km. The common altitude of commercial airliners is about 10 km.

— 99.99997% of the atmosphere by mass is below 100 km. The highest X-15 plane flight in 1963 reached an altitude of 354,300 ft (108,000 m).

Therefore, most of the atmosphere (99.9997%) is below 100 km, although in the rarefied region above this there are auroras and other atmospheric effects.

The mean molar mass of air is 28.97 g/mol. The composition figures above are by volume-fraction (V%), which for ideal gases is equal to mole-fraction (that is, fraction of total molecules). By contrast, *mass-fraction* abundances of gases, particularly for gases with significantly different molecular (molar) mass from that of air will differ from those by volume. For example, in air, helium is 5.2 ppm by *volume-fraction* and *mole-fraction*, but only about $(4/29) \times 5.2$ ppm = 0.72 ppm by *mass-fraction*.

Heterosphere

Below the turbopause at an altitude of about 100 km (not far from the mesopause), the Earth's atmosphere has a more-or-less uniform composition (apart from water vapor) as described above; this constitutes the homosphere.[6] However, above about 100 km, the Earth's atmosphere begins to have a composition which varies with altitude. This is essentially because, in the absence of mixing, the density of a gas falls off exponentially with increasing altitude but at a rate which depends on the molar mass. Thus higher mass constituents, such as oxygen and nitrogen, fall off more quickly than lighter constituents such as helium, molecular hydrogen, and atomic hydrogen. Thus there is a layer, called the heterosphere, in which the earth's atmosphere has varying composition. As the altitude increases, the atmosphere is dominated successively by helium, molecular hydrogen, and atomic hydrogen. The precise altitude of the heterosphere and the layers it contains varies significantly with temperature.

In pre-history, the Sun's radiation caused a loss of the hydrogen, helium and other hydrogen-containing gases from early Earth, and Earth was devoid of an atmosphere. The first atmosphere was formed by outgassing of gases trapped in the interior of the early Earth, which still goes on today in volcanoes.

Density and mass The density of air at sea level is about 1.2 kg/m^3(1.2 g/L). Natural variations of the barometric pressure occur at any one altitude as a consequence of weather. This variation is relatively small for inhabited altitudes but much more pronounced in the outer atmosphere and space because of variable solar radiation.

The atmospheric density decreases as the altitude increases. This variation can be approximately modeled using the barometric formula. More sophisticated models are used by meteorologists and space agencies to predict weather and orbital decay of satellites.

The average mass of the atmosphere is about 5 quadrillion metric tons or 1/1,200,000 the mass of Earth. According to the National Center for Atmospheric Research, "The total mean mass of the atmosphere is 5.1480×10^{18} kg with an annual range due to water vapor of 1.2 or 1.5×10^{15} kg depending on whether surface pressure or water vapor data are used; somewhat smaller than the previous estimate. The mean mass of water vapor is estimated as 1.27×10^{16} kg and the dry air mass as $5.1352 \pm 0.0003 \times 10^{18}$ kg."

Low Capacity

The atmosphere has "windows" of low opacity, allowing the transmission of electromagnetic radiation. The optical window runs from around 300 nanometers at roughly 400-700 nm, and continues up through the visual infrared to around 1100 nm, which is thermal infrared. There are also infrared and radio windows that transmit some infrared and radio waves. The radio window runs from about one centimeter to about eleven-meter waves

Importance of Atmosphere

The history of the Earth's atmosphere prior to one billion years ago is poorly understood and an active area of scientific research. The following discussion presents a plausible scenario.

The modern atmosphere is sometimes referred to as Earth's "third atmosphere", in order to distinguish the current chemical composition from two notably different previous compositions. The original atmosphere was primarily helium and hydrogen. Heat from the still-molten crust, and the sun, plus a probably enhanced solar wind, dissipated this atmosphere.

About 4.4 billion years ago, the surface had cooled enough to form a crust, still heavily populated with volcanoes which released steam, carbon dioxide, and ammonia. This led to the early "second atmosphere", which was primarily carbon dioxide and water vapor, with some nitrogen but virtually no oxygen. This second atmosphere had approximately 100 times as much gas as the current atmosphere, but as it cooled much of the carbon dioxide was dissolved in the seas and precipitated out as carbonates. The later "second atmosphere" contained largely nitrogen and carbon dioxide. However, simulations run at the University of Waterloo and University of Colorado in 2005 suggest that it may have had up to 40% hydrogen. It is generally believed that the greenhouse effect, caused by high levels of carbon dioxide and methane, kept the Earth from freezing.

One of the earliest types of bacteria was the cyanobacteria. Fossil evidence indicates that bacteria shaped like these existed approximately 3.3 billion years ago and were the first oxygen-producing evolving phototropic organisms. They were responsible for the initial conversion of the earth's atmosphere from an anoxic state to an oxic state (that is, from a state without oxygen to a state with oxygen) during the period 2.7 to 2.2 billion years ago. Being the first to carry out oxygenic photosynthesis, they were able to produce oxygen while sequestering carbon dioxide in organic molecules, playing a major role in oxygenating the atmosphere.

Photosynthesising plants later evolved and continued releasing oxygen and sequestering carbon dioxide. Over time, excess carbon became locked in fossil fuels, sedimentary rocks (notably limestone), and animal shells. As oxygen was released, it reacted with ammonia to release nitrogen; in addition, bacteria would also convert ammonia into nitrogen. But most of the nitrogen currently present in the atmosphere results from sunlight-powered photolysis of ammonia released steadily over the aeons from volcanoes.

As more plants appeared, the levels of oxygen increased significantly, while carbon dioxide levels dropped. At first the oxygen combined with various elements (such as iron), but eventually oxygen accumulated in the atmosphere, resulting in mass extinctions and further evolution. With the appearance of an ozone layer (ozone is an allotrope of oxygen) lifeforms were better protected from ultraviolet radiation. This oxygen-nitrogen atmosphere is the "third atmosphere". Between 200 and 250 million years ago, up to 35% of the atmosphere was oxygen (as found in bubbles of ancient atmosphere preserved in amber).

This modern atmosphere has a composition which is enforced by oceanic blue-green algae as well as geological processes. O_2 does not remain naturally free in an atmosphere but tends to be consumed (by inorganic chemical reactions, and by animals, bacteria, and even land plants at night), and CO_2 tends to be produced by respiration and decomposition and oxidation of organic matter. Oxygen would vanish within a few million years by chemical reactions, and CO_2 dissolves easily in water and would be gone in millennia if not replaced. Both are maintained by biological productivity and geological forces seemingly working hand-in-hand to maintain reasonably steady levels over millions of years.

Causes of Atmospheric Pollution

Motor vehicle emissions are likely the leading cause of air pollution. China, United States, Russia, Mexico, and Japan are the world leaders in air pollution emissions; however, Canada is the number two country, ranked per capita. Principal stationary pollution sources include chemical plants, coal-fired power plants, oil refineries, petrochemical plants, nuclear waste disposal activity, incinerators, large animal farms, PVC factories, metals production factories, plastics factories, and other heavy industry. Some of the more common soil contaminants are chlorinated hydrocarbons (CFH), heavy metals (such as chromium, cadmium—found in rechargeable batteries, and lead—found in lead paint, aviation fuel and still in some countries, gasoline), MTBE, zinc, arsenic and benzene. Ordinary municipal landfills are the source of many chemical substances entering the soil environment (and often groundwater), emanating from the wide variety of refuse accepted, especially substances illegally discarded there, or from pre-1970 landfills that may have been subject to little control in the U.S. or EU.

Pollution can also be the consequence of a natural disaster. For example hurricanes often involve water contamination from sewage, and petrochemical spills from ruptured boats or automobiles. Larger scale and environmental damage is not uncommon when coastal oil rigs or refineries are involved. Some sources of pollution, such as nuclear power plants or oil tankers, can produce widespread and potentially hazardous releases when accidents occur. In the case of noise pollution the dominant source class is the motor vehicle, producing about ninety percent of all unwanted noise worldwide.

Forms of Atmospheric Pollution

Although atmospheric pollution can have natural sources, for example volcanic eruptions, the term is usually used to refer to the gaseous by-products of man-made processes such as energy production, waste incineration, transport, deforestation and agriculture. During the last 200 years, mankind has begun to significantly alter the composition of the atmosphere through pollution. Although is still made up mostly of and , mankind, through its pollution, has increased the levels of many trace gases, and in some cases, released completely new gases to the atmosphere. Some of these trace gases, present in elevated concentrations, can be harmful to both humans and the environment.

Air pollution can result in poor air quality, both in cities and the countryside. Some air pollutants make people sick, causing breathing problems and increasing the likelihood of cancer. Others are harmful to plants, animals, and the ecosystems in which they live. Some air pollutants return to Earth in the form of acid rain, which corrodes statues and buildings, damages crops and forests, and makes lakes and streams unsuitable for fish and other plant and animal life.

Man-made air pollution is also changing the Earth's atmosphere so that it lets in more harmful radiation from the Sun. Although we have now banned products which can harm the Earth's , over Antarctica and the Arctic still form every year. At the same time, mankind is releasing more greenhouse gases to the atmosphere, preventing heat from escaping back into space and leading to a rise in global average temperatures. will raise sea levels and change climates all over the world. Some places will become hotter and drier, others wetter. The incidence of severe storms and flooding is likely to increase. Global warming will also affect food supply and increase the spread of tropical disease.

The major forms of pollution include:

- *Air pollution*, the release of chemicals and particulates into the atmosphere. Common examples include carbon monoxide, sulfur dioxide, chlorofluorocarbons (CFCs), and nitrogen oxides produced by industry and motor vehicles. Ozone and smog are created as nitrogen oxides and hydrocarbons react to sunlight.
- *Water pollution* via surface runoff and leaching to groundwater.

— *Noise pollution*, which encompasses roadway noise, aircraft noise, industrial noise as well as high-intensity sonar.

— *Soil contamination* occurs when chemicals are released by spill or underground storage tank leakage. Among the most significant soil contaminants are hydrocarbons, heavy metals, MTBE, herbicides, pesticides and chlorinated hydrocarbons.

— *Radioactive contamination*, added in the wake of 20th-century discoveries in atomic physics.

— *Light pollution*, includes light trespass, over-illumination and astronomical interference.

— *Visual pollution*, which can refer to the presence of overhead power lines, motorway billboards, scarred landforms, open storage of trash or municipal solid waste.

— *Thermal pollution*, is a temperature change in natural water bodies caused by human influence.

Air Pollution

Air pollution is a chemical, physical or biological agent that modifies the natural characteristics of the atmosphere. The atmosphere is a complex, dynamic natural gaseous system that is essential to support life on planet earth. Stratospheric ozone depletion due to air pollution has long been recognised as a threat to human health as well as to the earth's ecosystems. Worldwide air pollution is responsible for large numbers of deaths and cases of respiratory disease. Enforced air quality standards, like the Clean Air Act in the United States, have reduced the presence of some pollutants. While major stationary sources are often identified with air pollution, the greatest source of emissions is actually mobile sources, principally the automobile. Gases such as carbon dioxide, which contribute to global warming, have recently gained recognition as pollutants by some scientists. Others recognise the gas as being essential to life, and therefore incapable of being classed as a pollutant.

Air pollutants are classified as either directly released or formed by subsequent chemical reactions. A direct release air pollutant is one that is emitted directly from a given source, such as the carbon monoxide or sulfur dioxide, all of which are byproducts of combustion; whereas, a subsequent air pollutant is formed in the atmosphere through chemical reactions involving direct release pollutants. The formation of ozone in photochemical smog is the most important example of a subsequent air pollutant.

The World Health Organisation (WHO) estimates that 4.6 million people die each year from causes directly attributable to air pollution. Many of these mortalities are attributable to indoor air pollution. Worldwide more deaths per year are linked to air pollution than to automobile accidents. Research published in 2005 suggests that 310,000 Europeans die from air pollution annually. Direct causes of air pollution related deaths include aggravated asthma, bronchitis, emphysema, lung and heart diseases, and respiratory allergies. The US EPA estimates that a proposed set of changes in diesel engine technology could result in 12,000 fewer premature mortalities, 15,000 fewer heart attacks, 6,000 fewer emergency room visits by children with asthma, and 8,900 fewer respiratory-related hospital admissions each year in the United States.

The worst short term civilian pollution crisis in India was the 1984 Bhopal Disaster. Leaked industrial vapors from the Union Carbide factory, belonging to Union Carbide, Inc., U.S.A., killed more than 2,000 people outright and injured anywhere from 150,000 to 600,000 others, some 6,000 of whom would later die from their injuries. The United Kingdom suffered its worst air pollution event when the December 4th Great Smog of 1952 formed over London. In six days more than 4,000 died, and 8,000 more died within the following months. An accidental leak of anthrax spores from a biological warfare laboratory in the former USSR in 1979 near Sverdlovsk is believed to have been the cause of hundreds of civilian deaths.

Anthropogenic sources related to burning different kinds of fuel:

— Combustion-fired power plants

— Controlled burn practices used in agriculture and forestry management

— Motor vehicles generating air pollution emissions.

— Marine vessels, such as container ships or cruise ships, and related port air emissions

— Burning fossil fuels

— Burning wood, fireplaces, stoves, furnaces and incinerators

Other anthropogenic sources:

— Oil refining, power plant operation and industrial activity in general.

— Chemicals, dust and crop waste burning in farming.

- Fumes from paint, varnish, aerosol sprays and other solvents.
- Waste deposition in landfills, which generate methane.
- Military uses, such as nuclear weapons, toxic gases, germ warfare and rocketry.

Natural Sources:

- Dust from natural sources, usually large areas of land with little or no vegetation.
- Methane, emitted by the digestion of food by animals, for example cattle.
- Pine trees, which emit volatile organic compounds (VOCs).
- Radon gas from radioactive decay within the Earth's crust.
- Smoke and carbon monoxide from wildfires.
- Volcanic activity, which produce sulfur, chlorine, and ash particulates.

Between 1951 and 1991, the urban population has tripled, from 62.4 million to 217.6 million, and its proportion has increased from 17.3% to 25.7%. Nearly two-thirds of the urban population is concentrated in 317 class I cities, half of which lives in 23 metropolitan areas with populations exceeding 1 million. The number of urban agglomerations/cities with populations of over a million has increased from 5 in 1951 to 9 in 1971 and 23 in 1991.This rapid increase in urban population has resulted in unplanned urban development, increase in consumption patterns and higher demands for transport, energy, other infrastructure, thereby leading to pollution problems.

Growing Number of Vehicles

The number of motor vehicles has increased from 0.3 million in 1951 to 37.2 million in 1997. Out of these, 32% are concentrated in 23 metropolitan cities. Delhi itself accounts for about 8% of the total registered vehicles and has more registered vehicles than those in the other three metros (Mumbai, Calcutta, and Chennai) taken together. At the all-India level, the percentage of two-wheeled vehicles in the total number of motor vehicles increased from 9% in 1951 to 69% in 1997, and the share of buses declined from 11% to 1.3% during the same period. This clearly points to a tremendous increase in the share of personal transport vehicles.

In 1997, personal transport vehicles constituted 78.5% of the total number of registered vehicles. Road-based passenger transport has recorded very high growth in recent years especially since 1980-81. It is estimated that the roads accounted for 44.8 billion passenger kilometer in 1951 which has since grown to 2,515 billion PKM in 1996. The freight traffic handled by road in 1996 was about 720 billion tonne kilometer (TKM) which has increased from 12.1 TKM in 1951. In contrast, the total road network has increased only 8 times from 0.4 million kms in 1950-51 to 3.3 million kms in 1995-96. The slow growth of road infrastructure and high growth of transport performance and number of vehicles all imply that Indian roads are reaching a saturation point in utilising the existing capacities. The consumption of gasoline and HSD has grown more than 3 times during the period 1980-1997. While the consumption of gasoline and HSD were 1,522 and 9,050 thousand tonnes in 1980-81, it increased to 4,955 and 30,357 thousand tonnes in 1996-97, respectively.

Rapid Industrialisation

India has made rapid strides in industrialisation, and it is one of the ten most industrialised nations of the world. But this status, has brought with it unwanted and unanticipated consequences such as unplanned urbanisation, pollution and the risk of accidents.The CPCB (Central Pollution Control Board) has identified seventeen categories of industries (large and medium scale) as significantly polluting and the list includes highly air polluting industries such as integrated iron and steel, thermal power plants, copper/ zinc/ aluminium smelters, cement, oil refineries, petrochemicals, pesticides and fertiliser units.

The state-wise distribution of these pre- 1991 industries indicates that the states of Maharashtra, Uttar Pradesh, Gujarat, Andhra Pradesh and Tamil Nadu have a large number of industries in these sectors. The categorywise distribution of these units reveals that sugar sector has the maximum number of industries, followed by pharmaceuticals, distillery, cement and fertiliser. It also indicates that agro-based and chemical industries have major shares of 47% and 37% of the total number of industries respectively. The status of pollution control as on 30 June 2000 is as follows: out of 1,551 industries, 1,324 have so far been provided the necessary pollution control facilities, 165 industries have been closed down and the remaining 62 industries are defaulters. It may be noted that in some of the key sectors such as iron and steel, 6 out of 8 units belong to the defaulters category in terms of having pollution control facilities to comply with the standards. On the other hand, cement, petrochemicals and oil refinery sectors do not have any defaulters.

Small scale industries are a special feature of the Indian economy and play an important role in pollution. India has over 3 million small scale units accounting for over 40 percent of the total industrial output in the country. In general, Indian small scale industries lack pollution control mechanisms. While the larger industries are better organised to adopt pollution control measures, the small scale sector is poorly equipped to handle this problem. They have a very high aggregate pollution potential. Also, in many urban centres, industrial units are located in densely populated areas, thereby affecting a large number of people.

Power Consumption

Since 1950-51, the electricity generation capacity in India has multiplied 55 times from a meager 1.7 thousand MW to 93.3 thousand MW. The generating capacity in India comprises a mix of hydro, thermal, and nuclear plants. Since the early seventies, the hydro-thermal capacity mix has changed significantly with the share of hydro in total capacity declining from 43% in 1970-71 to 24% in 1998-99. Thermal power constitutes about 74% of the total installed power generation capacity. However, increasing reliance on this source of energy leads to many environmental problems. India's coal has a very high in ash content (24%-45%). The increased dependence of the power sector on an inferior quality coal has been associated with emissions from power plants in the form of particulate matter, toxic elements, fly ash, oxides of nitrogen, sulphur and carbon besides ash, which required vast stretches of land for disposal. During 1998-99, the power stations consumed 208 million tonnes of coal, which in turn produced 80 million tonnes of ash posing a major problem disposal. Thermal power plants belong to the 17 categories of highly polluting industries. As on 30 June 2000, out of the 97 pre-1991 TPP's, 20 plants had not yet provided the requisite pollution control facilities.

Domestic Pollution

Pollution from different types of cooking stoves using coal, fuelwood, and other biomass fuels contributes to some extent, to the overall pollution load in urban areas. For example, in Delhi, the share of the domestic sector is about 7%-8% of the total pollution load. The main concern is the use of inefficient and highly polluting fuels in the poorer households leading to a deterioration in indoor air-quality and health. However, a positive development in the domestic energy consumption is that liquefied petroleum gas is fast becoming the most popular cooking fuel, especially in urban areas, as it is cleaner and more efficient than traditional cooking fuels.

Other Pollutants

The problem of air pollution in urban areas is also aggravated due to inadequate power supply for industrial, commercial and residential activities due to, which consumers have to use diesel-based captive power generation units emitting high levels NO_X and SO_X. In addition, non-point sources such as waste burning, construction activities, and roadside air borne dust due to vehicular movement also contribute to the total emission load.

Air Pollutant Emission Load

The direct impact of a growth in various causal factors/pressures is the increase in the emission loads of various pollutants, which has led to deterioration in the air quality. In India, there is no systemic time series data available related to air pollutant emission loads and trends. The availability of emission factors for Indian conditions is another issue that has not been given due attention so far. TERI provides some broad estimates of the increase in pollution load from various sectors in India. The total estimated pollution load from the transport sector increased from 0.15 million tonnes in 1947 to 10.3 million tonnes in 1997.

In 1997, CO claimed the largest share (43%) of the total, followed by NO_X (30%), HC (20%), SPM (5%), and SO_2 (2%). Likewise, in the thermal power sector, the total estimated pollution load of SPM, SO_2 and NO_X increased from 0.3 million tonnes in 1947 to 15 million tonnes in 1997. In 1997, SPM claimed the largest share (86%) of the total. In the industrial sector, the total estimated emissions of SPM from the 7 critical industries (iron and steel, cement, sugar, fertilisers, paper and paper board, copper and aluminium) increased from 0.2 million tonnes in 1947 to 3 million tonnes in 1997.

The World Bank study shows that pollution is concentrated among a few industrial sub-sectors and that a sector's contribution to pollution is often disproportionate to its contribution to industrial output. For example, petroleum refineries, textiles, pulp and paper, and industrial chemicals produce 27% of the industrial output but contribute 87% of sulphur emissions and 70% of nitrogen emissions from the industrial sector.The drastic increase in number of vehicles has also resulted in a significant increase in the emission load of various pollutants.The quantum of vehicular pollutants emitted is highest in Delhi followed by Mumbai, Bangalore, Calcutta and Ahmedabad. Carbon monoxide (CO) and hydrocarbons (HC) account for 64% and 23%, respectively, of the total emission load due to vehicles in all these cities considered together.

Apart from the concentration of vehicles in urban areas, other reasons for increasing vehicular pollution are the types of engines used, age of vehicles, congested traffic, poor road conditions, and outdated automotive technologies and traffic management systems. Vehicles are a major source of pollutants in metropolitan cities. In Delhi, the daily pollution load has increased from 1,450 tonnes in 1991 to 3,000 metric tonnes in 1997. The share of the transport sector has increased from 64% to 67% during the same period while that of the industrial sector (including power plants) has decreased from 29% to 25%.

Suspended Particulate Matter (SPM)

Suspended particulate matter is one of the most critical air pollutants in most of the urban areas in the country and permissible standards are frequently violated several moni-tored locations. Its levels have been consistently high in various cities over the past several years. The mean of average values of SPM for nine years (1990 to 1998) ranged between 99 $\mu g/m^3$ and 390 $\mu g/m^3$ in residential areas and between 123 $\mu g/m^3$ and 457 $\mu g/m^3$ in industrial areas indicating that the annual average limit of suspended particulate matter for residential areas (140 $\mu g/m^3$) and for industrial areas (360 $\mu g/m^3$) had been frequently violated in most cities.

The maximum suspended particulate matter (SPM) values were observed in Kanpur, Calcutta, and Delhi, while low values have been recorded in the south Indian cities of Chennai, Bangalore, and Hyderabad. The SPM non-attainment areas are dispersed throughout the country. The states with maximum SPM problems are Gujarat, Maharashtra, and Madhya Pradesh, where SPM problems are high to critical in a large number of cities. The widespread criticality of the SPM problem in the country is due to the synergistic effects of both anthropogenic and natural sources. Some of these are extensive urbanisation and construction activities, vehicular pollution increase, extensive use of fossil fuel in industrial activities, inadequacy of pollution control measures, biomass burning, presence of large acid and semi-acid area in north-west part of India, increasing desertification, and decreasing vegetation cover.

SO_2

The annual average level fluctuation of SO_2 was highest in residential areas of Howrah (West Bengal) recording between 40.6 $\mu g/m^3$ and 103.8 $\mu g/m^3$ while it was quite low in residential areas of Nagpur, Chandigarh, and Jaipur. Among the industrial areas, the recorded sulphur dioxide levels were high at Pondicherry, Calcutta, Mumbai, and Howrah, and low at Nagpur, Jaipur, and Chandigarh. Thus, based on the mean average sulphur dioxide value, Nagpur, Chandigarh and Jaipur are cities with the least problems related to sulphur dioxide in the ambient air, while the problem is significant in Howrah, Calcutta, and Pondicherry, where annual average limits have been violated many times during the past several years.

The sulphur dioxide levels have generally attained air quality standards in the country except some cities of dense urban and industrial activities like Dhanbad (Bihar), Ahmedabad, Ankleshwar, Vadodara and Surat (Gujarat); Nagda (Madhya Pradesh); Pondicherry; and Howrah (West Bengal). Some of the measures taken such as cleaner fuel quality and switch over to cleaner fuel option have contributed t lower SO_2 ambient levels.

NO_2

The air quality monitoring data indicate that the annual average nitrogen dioxide has been well within the annual average limit (60 $\mu g/m^3$ for residential area and 80 $\mu g/m^3$ for industrial areas) at most urban cities except in some years in residential areas of Howrah, Vishakhapatnam, Kota, and industrial areas of Howrah. The annual average concentration has been low at Nagpur, Chennai, Kanpur, and Chandigarh while levels are moderate in other cities.The nitrogen dioxide non-attainment areas were at Vishakhapatnam (AP), Jabalpur (MP), Pondicherry, Alwar, Kota, Udaipur (Rajasthan) and Howrah (West Bengal). The criticality of problem was observed at Vishakhapatnam, Kota, and Howrah.

Air Quality at Traffic Intersections

Air quality monitoring conducted at different traffic intersections in Delhi revealed the following:

- Respirable particulate matter was excessively high at all the monitoring locations.
- Sulphur dioxide was recorded within limits at all the locations.

— Nitrogen dioxide was recorded well within the limits except a few locations.

— The carbon monoxide levels at most locations was much higher than the prescribed permissible limit. This is because of high traffic density and large number of motor vehicles operating on the roads.

Health Impacts

In India, millions of people breathe air with high concentrations of dreaded pollutants. The air is highly polluted in terms of suspended particulate matter in most cities. This has led to a greater incidence of associated health effects on the population manifested in the form of sub-clinical effects, impaired pulmonary functions, use of medication, reduced physical performance, frequent medical consultations and hospital admissions with complicated morbidity and even death in the exposed population. As per a World Bank study, respiratory infections contribute to 10.9% of the total burden of diseases, which may be both due to presence of communicable diseases as well as high air pollution levels, while cerebro vascular disease (2.1%) ischemic heart disease (2.8%) and pulmonary obstructions (0.6%) are much lower. The prevalence of cancer is about 4.1% amongst all the diseases indicating that the effects of air pollution are visualised on the urban population.

A WHO/UNEP study compared standardised prevalences of respiratory diseases in different areas of Mumbai, classified according to ambient average concentrations of sulphur dioxide. The study revealed a relatively higher prevalence of most respiratory diseases in polluted urban areas than in the rural control area. In India, in a study of 2031 children and adults in five mega cities, of the 1852 children tested, 51.4% had blood lead levels above 10 μg/dl. The percentage of children having 10 μg/dl or higher blood lead levels ranged from 39.9% in Bangalore to 61.8% in Mumbai. Among the adults, 40.2% had blood lead levels of about 10 μg/dl. Using the 1991-92 air pollution data for particulates, SO_2, NO_X, and lead from 36 cities, health impacts were estimated in terms of reductions in morbidity and mortality if pollutant levels in these cities were reduced to the WHO annual average standard. The total health costs due to air pollution were estimated to be $517-2102 million. Also, the physical impacts were in terms of 40,000 premature deaths avoided.

TERI estimated the incidence of mortality and morbidity in different groups in India due to exposure to PM10 and translated these impacts into economic values. The results indicated 2.5 million premature deaths and total morbidity and mortality costs of Rs 885 billion to Rs 4250 billion annually. Studies by the CPCB on the ambient noise levels show that they exceed the prescribed standards in most of the big cities. The major sources of noise are vehicles and industrial manufacturing processes.

Noise Pollution

Noise is an unwanted, unpleasant and annoying sound caused by vibration of the matter. Vibrations impinge on the ear drum of a human or animal and setup a nervous disturbance, which we call sound. When the effects of sound are undesirable that it may be termed as "Noise". Noise from industry, traffic, homes and recreation can cause annoyance, disturb sleep and affect health. Thus, sound is a potentially serious pollutant and threat to environmental health. The response of the human ear to sound depends both on the sound frequency (measure in Hertz, Hz) and the sound pressure, measured in decibels (dB). A normal ear in healthy young person can detect sounds with frequencies from 20Hz to 20,000 Hz. Noise measurements are expressed by the term Sound Pressure Level (SPL) which is logarithmic ratio of the sound pressure to a reference pressure and is expressed as a dimensionless unit of power, the decibel (dB). The reference level is 0.0002 microbars, the threshold of human hearing.

Decibel $L_{eq} = 10 \log 10 \; L/L_{o}$

L_{eq} = Equivalent Noise level, L = Sound Intensity, L_{o} = Reference level

If a measurement of noise emission is required a sound level meter is used. A measure of the level of sound is called the decibel. The zero of the decibel scale is the hearing threshold. Sounds at 0-10 decibel are so quiet that they are almost impossible to hear, while at the top end of the scale, at around 150 decibel, it can damage your eardrums.

Community Noise

Community noise (also called environmental noise, residential noise or domestic noise) is defined as noise emitted from all sources, except noise at the industrial workplace. Main sources of community noise include road, rail and air traffic, construction and public work, and the neighbourhood. Typical neighbourhood noise comes from premises and installations related to the catering trade (restaurant, cafeterias, discotheques, etc.); from live or recorded music; from sporting events including motor sports; from playgrounds and car parks; and from domestic animals such as barking dogs. The main indoor sources are ventilation systems, office machines, home appliances and neighbours.

Roadway noise

Roadway noise is the collective sound energy emanating from motor vehicles. In the USA it contributes more to environmental noise exposure than any other noise source, and is constituted chiefly of engine, tire, aerodynamic and braking elements.Roadway noise began to be measured in a widespread manner in the 1960s, when computer modeling of this phenomenon was perfected. After passage of the National Environmental Policy Act and Noise Control Act, the demand for detailed analysis soared, and decision makers began to look to acoustical scientists for answers regarding the planning of new roadways and the design of noise mitigation. The intensity of roadway noise is governed by the following variables: traffic operations (speed, truck mix, age of vehicle fleet), roadway surface type, tire types, roadway geometrics, terrain, micrometeorology and the geometry of area structures.

Traffic operations noise is affected significantly by vehicle speeds, since sound energy roughly doubles for each increment of ten miles an hour in vehicle velocity; an exception to this rule occurs at very low speeds where braking and acceleration noise dominate over aerodynamic noise. Small reductions in vehicle noise occurred in the 1970s as states and provinces enforced unmuffled vehicle ordinances. The vehicle fleet noise has not changed very much over the last three decades; however, if the trend in hybrid vehicle use continues, substantial noise reduction will occur, especially in the regime of traffic flow below 35 miles per hour. As a pedestrian safety issue, hybrid vehicles are so quiet at low speeds that the customary warning noise may not alert the pedestrian to nearby danger, creating a potential hazard for visually-impaired people, who rely on such noise to navigate in areas of heavy traffic, find it particularly difficult to, for instance, cross streets. Trucks contribute a disproportionate amount of noise not only because of their large engines, but also the height of the diesel stack and the aerodynamic drag. Significant interior noise is usually present inside moving motor vehicles; in fact, passengers are generally not aware that these levels are high, because experience has led motorists to expect levels commonly exceeding 65 dBA.

Roadway surface types contribute differential noise effects of up to 4 dB, with chip seal type and grooved roads being the loudest and concrete surfaces without spacers being the quietest. Asphaltic surfaces are about average.Tire types had considerable design changes in the 1970s, and at this juncture are probably optimised for noise control, given the of safety needs for a significant grip by the tread. Roadway geometrics and surrounding terrain are interrelated, since the propagation of sound is sensitive to the overall geometry and must consider diffraction (bending of sound waves around obstacles), reflection, ground wave attenuation, spreading loss and refraction. A simple discussion indicates that sound will be diminished when the path of sound is blocked by terrain, or will be enhance if the roadway is elevated so as to broadcast; however, the complexities of variable interaction are so great, that there are many exceptions to this simple argument.

Micrometeorology is significant in that sound waves can be refracted by wind gradients or thermoclines, effectively dismissing the effect of some sound barriers or terrain intervention. Geometry of area structures is an important input, since the presence of buildings or walls can block sound under certain circumstances, but reflective properties can augment sound energy at other locations. Because of the complexity of the variables discussed, it is necessary to create a computer model that can analyse sound levels in the vicinity of roadways. The first meaningful models arose in the late 1960s and early 1970s addressing the noise line source. Two of the leading research teams were BBN in Boston and ESL of Sunnyvale, California. Both of these groups developed complex mathematical models to allow the study of alternate roadway designs, traffic operations and noise mitigation strategies in an arbitrary setting. Later model alterations have come into widespread use among state Departments of Transportation and city planners, but the accuracy of early models has had little change in 40 years.

Generally the models trace sound ray bundles and calculate spreading loss along with ray bundle divergence (or convergence) from refractive phenomena. Diffraction is usually addressed by establishing secondary emitters at any points of topographic or anthropomorphic "sharpness" (such as noise barriers or building surfaces). Meteorology can be addressed in a statistical manner allowing for actual wind rose and wind speed statistics.

An interesting early case where two of the leading models were pitted against each other involved a proposed widening of the New Jersey Turnpike from six to twelve lanes. The BBN and ESL models were on opposing sides of a matter decided in New Jersey Superior Court. This case in the early 1970s was one of the first U.S. examples of acoustical scientists playing a role in the design of a major highway. The models allowed the court to understand the effects of roadway geometry (width in this case), vehicle speeds, proposed noise barriers, residential setback and pavement types. The outcome was a compromise that involved substantial mitigation of noise pollution impacts.

Another early case involved the proposed extension of Interstate 66 through Arlington, Virginia. The plaintiff, Arlington Coalition on Transportation sued the Virginia Department of Transportation on the grounds of air quality, noise and neighborhood disruption. To analyse roadway noise, the ESL model was used by the plaintiff, who won this case partially due to the credibility of the computer model. The matter was revisited a decade later and a greatly reduced highway design with transit element and extensive noise mitigation was agreed to. Later cases have occurred in every state, both in contentious actions and in routine highway planning and design. The public as well as governmental agencies have become aware of the value of acoustical science to provide useful insights to the roadway design process.

Aircraft noise

Aircraft noise is defined as sound produced by any aircraft on run-up, taxiing, take off, over-flying or landing. Aircraft noise is a significant concern for approximately 100 square kilometers surrounding most major airports. Aircraft noise is the second largest (after roadway noise) source of environmental noise. While commercial aviation produces the preponderance of total aircraft noise, private aviation and military operations also play a role. Take-off of aircraft may lead to a sound level of more than 100 dBA at the ground, with approach and landing creating lower levels. Since aircraft landing in inner-city airports are often lower than 60 meters above roof level, a sound level above 100 dBA can be realised.

A moving aircraft including the jet engine or propeller causes compression and rarefaction of the air, producing motion of air molecules. This movement propagates through the air as pressure waves. If these pressure waves are strong enough and within the audible frequency spectrum, a sensation of hearing is produced. Different aircraft types have different noise levels and frequencies. The contributions to the total noise level originate from three main sources:

— Aerodynamic noise
— Engine and other mechanical noise
— Noise from aircraft systems

Aerodynamic noise

Aerodynamic noise arises from the airflow around the aircraft fuselage and control surfaces. This type of noise increases with aircraft speed and also at low altitudes due to the density of the air. Jet-powered aircraft create intense noise from aerodynamics, which is typically broadband. Low flying, high speed military aircraft produce especially loud aerodynamic noise. The shape of the nose, windshield or canopy of an aircraft can greatly affect the sound produced. Much of the noise of a propeller aircraft is of aerodynamic origin due to the flow of air around the blades. The helicopter main and tail rotors also give rise to aerodynamic noise. This type of aerodynamic noise is mostly low frequency determined by the rotor speed.

Engine and other mechanical noise

Much of the noise in propeller aircraft comes equally from the propellers and aerodynamics. Helicopter noise has a unique spectral content, essentially being aerodynamically induced noise from the main and tail rotors and mechanically induced noise from the main gearbox and various transmission chains. The mechanical sources produce narrow band high intensity peaks relating to the rotational speed and movement of the moving parts. In computer modelling terms noise from a moving aircraft can be treated as a line source.

Noise from aircraft systems

Cockpit and cabin pressurisation and conditioning systems are often a major contributor within cabins of both civilian and military aircraft. However, one of the most significant sources of cabin noise from commercial jet aircraft other than the engines is the Auxiliary Power Unit (or APU). An Auxiliary Power Unit is a relatively small self contained generator used in aircraft to start the main engines, usually with compressed air, and to provide electrical power while the aircraft is on the ground. The typical noise output of an APU is 113 decibels. This is about 27 decibels lower than that of a jet engine. Other internal aircraft systems can also contribute, such as specialised electronic equipment in some military aircraft.Lesser intensities of noise are produced for cruising velocities, mainly due to the altitudes of operation. However, this noise often is heard in country settings which are by nature very peaceful. Thus the intrusion of this type of noise can be very intrusive even if much less in amplitude. Landing aircraft descend on a three degree glide path towards an aiming point approximately 300 meters from the runway threshold. This places them at 60 meters above the ground at about 1200 meters from the aiming point or 900 meters from the start of the runway. This distance is usually outside the airport fence. Departing aircraft normally are over 150 meters above the ground before crossing the end of the runway.

Health effects of aircraft noise

The annoyance effects of aircraft noise are widely recognised; however, aircraft noise is also responsible for a significant amount of hearing loss as well as a contributor to a number of diseases. Only in the early 1970s did aircraft noise become a widespread topic of concern in the U.S. and federal regulations began to recognise the significance of abating these impacts in the vicinity of major commercial airports. High levels of aircraft noise that commonly exist near major commercial airports are known to increase blood pressure and contribute to hearing loss. Some research indicates that it contributes to heart diseases, immune deficiencies, neurodermatitis, asthma and other stress related diseases. Further research is being carried out to better understand these effects.

Prior research indicates clearly that hearing loss is less a product of aging than a result of exposure to transportation related noise. Any sound louder than normal conversation can damage the delicate hair cells in the cochlea, the structure in the inner ear that converts sound waves into auditory nerve signals. Initially damage to the cochlea may be temporary, but with repeated exposure, the damage becomes permanent and tinnitus maybe develop. More recently the Centers for Disease Control and Prevention's (CDC) National Center for Environmental Health (NCEH) conducted an analysis to determine the prevalence of hearing loss among children using data collected from 1988-1994 in the Third National Health and Nutrition Examination Survey. The analysis indicates that 14.9% of U.S. children have low or high frequency hearing loss of at least 16 dB hearing level in one or both ears. From research of the National Institutes of Health, roughly 65 million Americans are exposed to sound levels that can interfere with their function at work or disrupt sleep, and 25 million are exposed to health risk (cardiovascular, immunological, etc.) from environmental noise.

Occupational Noise

The many and varied sources of noise in industrial machinery and processes include: rotors, gears, turbulent fluid flow, impact processes, electrical machines, internal combustion engines, pneumatic equipment, drilling, crushing, blasting, pumps and compressors. Furthermore, the emitted sounds are reflected from floors, ceiling and equipment. Noise is a common occupational hazard in many workplaces. Occupational exposure limits specify the maximum sound pressure levels and exposure times to which nearly all workers may be repeatedly exposed without adverse effect on their ability to hear and understand normal speech. Occupational noise can be two of types:

— *Continous noise.* It is produced by machinery that operates without interruption in the same mode e.g. blowers, pumps and processing equipment.

— *Intermittent noise.* When machinery operates in cycles or when single vehicles or aeroplanes pass by the noise level increases and decreases rapidly.

Urbanisation and Noise Pollution

Urbanisation, ie., the growth of urban population, has awfully accelerated during this century and has been faster in developing countries than in the advanced ones. The problems associated with the rapid change in human environment have intensified in the cities. It is here that the physical environment is becoming increasingly polluted, the man-made environment of slums, restricted living space and noise are at their worst and the changes in the social environments have aggravated many problems. Mankind is constantly flowing from rural settlements and every proposal for stopping this flow is completely unrealistic. Scientific and technological developments have given momentum to the growth of industries and hence, the pace of urbanisation. Man has been living on the Earth, for about 40,000 years. He is surrounded by various forms of 'Organisms' 'forces' and 'conditions'—both physical and biological, eg. sunlight, land, air water and living beings, which include all types of plants and Animals. The total of these is called Environment. For the first time in his entire cultural history man has been confronted with the most horrible, tragic and unprecedented problems of "Environment Pollution". Not very far back, in the past, this very environment was pure, virgin and uncontaminated, and basically quite hospitable for mankind. It is all due to thoughtless over-exploitation of our various natural resources, by our own activities, perhaps due to our unending greed, in the garb of 'development', and the egoistic attitude towards "Nature" .The other three main reasons are:

(i) Population explosion

(ii) Rapid urbanisation and

(iii) The throw-away concept of disposable items.

Noise pollution did not create much public concern due to ignorance about the treacherous effect of noise on both workers in industry in particular and the public in the community in general. It is, therefore, imperative to assess the environment in which the noise is being heard by using suitable bases of judgement and awareness to determine whether or not a definite nuisance exists. Noise is an important environmental pollutant like noxious gases that befoul our air, water and soil. It destroys bridges and produces cracks in buildings. The noise can cause skin and mental diseases.

It has been revealed that noise is a technology generated problem and that the overall noise doubles every ten years keeping pace with our social and industrial progress. This geometric progression-wise growth of noise could be mind-boggling in view of the ever increasing pace of technological growth.

Noise is a major factor of environmental pollution; on the one hand, industrialisation, scientific and technological developments have contributed a great deal to the progress of society, on the other, these are main causes of environmental pollution, including noise pollution. As the day rises, the noise level in the different parts of the city increase in and around work places and homes. The peak noise levels are reached in the twilight hours as traffic reaches a peak. In India, the problem caused by noise pollution is more aggravated in view of the fact that there is hardly any celebration, festival, marriage or religious function, where there is no use of loud speakers at a very high pitch continuously for a long time. In offices also there is noise pollution due to clicking of typewriters, bells, ringing telephones, clattering office machines and conversations. On the road, noise pollution exists due to growing automobiles, screeching tyres, squealing brakes, screaming sirens, blaring televisions and radios, and blasting horns. Another major factor contributing to the noise pollution in India is that in many of the cities, the industrial and commercial units are either not very far from the residential areas, or they are sometimes set up in the residential areas.

A vibrating source produces vibration into the medium in which it is placed. These vibrations are propagated as waves in the form of pressure variations and are termed as acoustic waves. If they fall within the range capable of exciting the sense of hearing, they are called as sound waves. An acoustic waves travels in a given medium at a constant velocity. When the level of the sound becomes objectionable, it is called noise. Thus in general, the sound can be referred as a physical or mechanical disturbance capable of being detected by the human ear. The human ear can detect sounds from 20 Hz to 20,000 Hz. The frequencies most important for understanding normal speech like between 300 Hz. to 5,000 Hz.

Effects of Noise on Human Beings

(a) *Auditory effect*: The human ear is a very sensitive instrument. If the hearing mechanisms are damaged in any way either by excessive noise levels or by diseases which affect the brain, the auditory nerve or the auditory ossicles, then hearing will be impaired. Intense noise levels, for example, are encountered in many industries and they can cause temporary or progressively permanent loss of hearing.

(b) *Physiological effect*: Noise is likely to harm the physiological and psychological well-being of people. Physiological changes include:

(c) *Non-auditory effects:* In addition to its auditory effects (temporary and permanent threshold shifts, noise inducted deafness etc.), noise can also produce many nonauditory effects. The exposure to noise.

— May interfere with verbal communication

— Cause annoyance and distraction

— Reduce working efficiency and work output

— Cause fatigue.

Sleep interference: The WHO- World Health Organisation Task Group on Environmental Health Criteria for Noise has recommended a noise level of 35 (A) to preserve the restorative process of sleep.

Speech interference: Speech reception is the most important and also the most complex use of the auditory system. Noise can either mask the speech to make it inaudible or by making only some frequencies leaving it audible but of reduced intelligibility.

A decibel can be defined as an abstract unit. It is to be remembered that the threshold at the normal hearing is 20-25 decibels and of normal conversation is 60 decibels. It has also been noticed that speech interference occurs at 75 decibels and definite annoyance begins at 80 decibels. The motor- activities disturbed at 90 decibels and physiological disturbance occurs beyond 120 decibels. Definite pain occurs at 140 decibels and though no human being has reportedly died of noise, experimental results have shown that mice died at 175 decibels of noise. According to environmentalist Thomas G. Aylesworth, "Constant noise may cause our blood-vessels to contract, our skin to become a pale, our muscles to construct and adrenaline to be short into our blood-stream". This is the reason why factory workers develop abnormal heartbeat rates and suffering from insomania, nervousness and impaired motor coordination.

Noise Pollution Control

In the days before the development of environmental jurisprudence, the Common Law remedy against nuisance was the only means available to curtail excessive noise, and this was wholly based on the discretion of the judge. Whether a particular noise constitutes a nuisance, after all, is often a question of degree. The development of standards for public conduct, as well as case law in this area, has led to significant changes in our understanding of noise pollution and the measures adopted to check it. Although a regulatory environment has slowly been built up around many activities, these do not usually address noise pollution specifically. Even when regard for the public is taken into consideration, the laws usually confine themselves to other matters, or do not adequately address noise issues. For example:

— *The Factories Act, 1948*: Surprisingly, no industrial law provides any protection to workers from noise pollution. Section 11 (I) of the Act stipulates that every factory shall be kept clean, without having any nuisance; the word 'nuisance' may include noise, but statutory provisions in industrial law to provide this protection specifically are overdue. It is also noteworthy that under section 35 of the Act, protection is given for the eyes of an employee but no such protection is provided to ears.

— *Motor Vehicles*: Vehicles are among the chief noise producers in modern times. The Motor Vehicles Act, 1988 through sections 20, 21 (j), 41, 68, 68 I, 70, 91 and 111 empowers a state government to frame rules for the upkeep of motor vehicles and control of noise produced by them. If the provisions of the Act were even half implemented, noise in urban areas would reduce considerably. But the Motor Vehicles Rules made by states do not contain any effective control measures to control noise pollution except a meager control of horns and silencers of the motor vehicles.

— *The Aircrafts Act, 1934*: The Central Government has the power to make rules for manufacture, possession, use, operation, sale, import or export any aircraft, as well as the regulation of air transport services. The Act has many provisions but none for the control of noise. In this regard it is suggested that aerodromes be constructed far away from the residential areas of a city in order to protect residents from the noise created by frequent take-offs and landings. But three of India's busiest airports are located close to residential localities.

In the absence of an adequate regulatory framework specific to noise pollution, the status quo has been determined partly by the interpretation of other laws. Important among these have been Article 19 of the Constitution, which guarantees the fundamental right to freedom of speech and expression, and Article 25, which protects the free profession of one's religion. The use of a loudspeaker, or setting of fire-crackers, has assumed the status of a fundamental right by virtue of these two articles. Municipal bye-laws regulating their use have been enacted, but must take care not to limit the freedoms afforded by the articles. Also, unless the connections between noise and health are first judicially established, prohibitions against their use are difficult to pass. The judiciary has nonetheless weighed in on questions of noise pollution.

Waste and Water Pollution

Water pollution has many sources. The most polluting of them are the city sewage and industrial waste discharged into the rivers. The facilities to treat waste water are not adequate in any city in India. Presently, only about 10% of the waste water generated is treated; the rest is discharged as it is into our water bodies. Due to this, pollutants enter groundwater, rivers, and other water bodies. Such water, which ultimately ends up in our households, is often highly contaminated and carries disease-causing microbes. Agricultural run-off, or the water from the fields that drains into rivers, is another major water pollutant as it contains fertilisers and pesticides.

During the last fifty years, the number of industries in India has grown rapidly. But water pollution is concentrated within a few subsectors, mainly in the form of toxic wastes and organic pollutants. Out of this a large portion can be traced to the processing of industrial chemicals and to the food products industry. In fact, a number of large- and medium-sized industries in the region covered by the Ganga Action Plan do not have adequate effluent treatment facilities. Most of these defaulting industries are sugar mills, distilleries, leather processing industries, and thermal power stations. Most major industries have treatment facilities for industrial effluents. But this is not the case with small-scale industries, which cannot afford enormous investments in pollution control equipment as their profit margin is very slender.

Chemical Pollution

As rapidly developing countries such as India industrialise, the dangers to local communities from pollution are often overlooked until there is a major disaster such as occurred in Bhopal.

The effects of chamical pollution is being rapidly felt across India. It has found that the incidence of diseases related to nervous, circulatory, respiratory, digestive and endocrine systems was one to four times higher in heavily industrialised areas as compared to unindustrialised areas. Many cases of congenital deformity and chromosomal abnormalities were also reported, in addition to 11 cases of different kinds of cancer. Skin disorders are also rampant.

The wave of industrialisation that began in the late 1970s has changed the complexion of India's once placid landscape. Lakes, streams, as well as the groundwater are laced with toxic heavy metals and chemicals, as proved by several studies by government agencies and research institutions including the National Geophysical Research Laboratory.

Reduction Efforts

There are various air pollution control technologies and land use planning strategies available to reduce air pollution. At its most basic level land use planning is likely to involve zoning and transport infrastructure planning.

In most developed countries, land use planning is an important part of social policy, ensuring that land is used efficiently for the benefit of the wider economy and population as well as to protect the environment.

Efforts to reduce pollution from mobile sources includes primary regulation (many developing countries have permissive regulations), expanding regulation to new sources (such as cruise and transport ships, farm equipment, and small gas-powered equipment such as lawn trimmers, chainsaws, and snowmobiles), increased fuel efficiency (such as through the use of hybrid vehicles), conversion to cleaner fuels (such as bioethanol, biodiesel, or conversion to electric vehicles).

Control Devices

The following items are commonly used as pollution control devices by industry or transportation devices. They can either destroy contaminants or remove them from an exhaust stream before it is emitted into the atmosphere.

— *Electrostatic precipitators.* An electrostatic precipitator (ESP), or electrostatic air cleaner is a particulate collection device that removes particles from a flowing gas (such as air) using the force of an induced electrostatic charge. Electrostatic precipitators are highly efficient filtration devices that minimally impede the flow of gases through the device, and can easily remove fine particulates such as dust and smoke from the air stream.

— *Baghouses.* Designed to handle heavy dust loads, a dust collector consists of a blower, dust filter, a filter-cleaning system, and a dust receptacle or dust removal system (distinguished from air cleaners which utilize disposable filters to remove the dust).

— *Particulate scrubbers.* Wet scrubber is a form of pollution control technology. The term describes a variety of devices that use pollutants from a furnace flue gas or from other gas streams. In a wet scrubber, the polluted gas stream is brought into contact with the scrubbing liquid, by spraying it with the liquid, by forcing it through a pool of liquid, or by some other contact method, so as to remove the pollutants.

Legal Regulations

In general, there are two types of air quality standards. The first class of standards (such as the U.S. National Ambient Air Quality Standards and E.U. Air Quality Directive) set maximum atmospheric concentrations for specific pollutants. Environmental agencies enact regulations which are intended to result in attainment of these target levels. The second class (such as the North American Air Quality Index) take the form of a scale with various thresholds, which is used to communicate to the public the relative risk of outdoor activity. The scale may or may not distinguish between different pollutants.

In Canada air pollution and associated health risks are measured with the The Air Quality Health Index or (AQHI). It is a health protection tool used to make decisions to reduce short-term exposure to air pollution by adjusting activity levels during increased levels of air pollution.

The Air Quality Health Index or "AQHI" is a federal program jointly coordinated by Health Canada and Environment Canada. However, the AQHI program would not be possible without the commitment and support of the provinces, municipalities and NGOs. From air quality monitoring to health risk communication and community engagement, local partners are responsible for the vast majority of work related to AQHI implementation. The AQHI provides a number from 1 to 10+ to indicate the level of health risk associated with local air quality. Occasionally, when the amount of air pollution is abnormally high, the number may exceed 10. The AQHI provides a local air quality current value as well as a local air quality maximums forecast for today, tonight and tomorrow and provides associated health advice.

As it is now known that even low levels of air pollution can trigger discomfort for the sensitive population, the index has been developed as a continuum: The higher the number, the greater the health risk and need to take precautions. The index describes the level of health risk associated with this number as 'low', 'moderate', 'high' or 'very high', and suggests steps that can be taken to reduce exposure.

It is measured based on the observed relationship of Nitrogen Dioxide (NO2), ground-level Ozone (O_3) and particulates (PM2.5) with mortality from an analysis of several Canadian cities. Significantly, all three of these pollutants can pose health risks, even at low levels of exposure, especially among those with pre-existing health problems.

When developing the AQHI, Health Canada's original analysis of health effects included five major air pollutants: particulates, ozone, and nitrogen dioxide (NO_2), as well as sulfur dioxide (SO_2), and carbon monoxide

(CO). The latter two pollutants provided little information in predicting health effects and were removed from the AQHI formulation.

In Europe, Council Directive 96/62/EC on ambient air quality assessment and management provides a common strategy against which member states can "set objectives for ambient air quality in order to avoid, prevent or reduce harmful effects on human health and the environment . . . and improve air quality where it is unsatisfactory".

On 25 July 2008 in the case Dieter Janecek v Freistaat Bayern CURIA, the European Court of Justice ruled that under this directive citizens have the right to require national authorities to implement a short term action plan that aims to maintain or achieve compliance to air quality limit values.

This important case law appears to confirm the role of the EC as centralised regulator to European nation-states as regards air pollution control. It places a supranational legal obligation on the UK to protect its citizens from dangerous levels of air pollution, furthermore superseding national interests with those of the citizen.

In 2010, the European Commission (EC) threatened the UK with legal action against the successive breaching of PM10 limit values. The UK government has identified that if fines are imposed, they could cost the nation upwards of £300 million per year.

In March 2011, the City of London remains the only UK region in breach of the EC's limit values, and has been given 3 months to implement an emergency action plan aimed at meeting the EU Air Quality Directive. The City of London has dangerous levels of PM10 concentrations, estimated to cause 3000 deaths per year within the city. As well as the threat of EU fines, in 2010 it was threatened with legal action for scrapping the western congestion charge zone, which is claimed to have led to an increase in air pollution levels.

In response to these charges, Boris Johnson, Mayor of London, has criticised the current need for European cities to communicate with Europe through their nation state's central government, arguing that in future "A great city like London" should be permitted to bypass its government and deal directly with the European Commission regarding its air quality action plan.

In part, this is an attempt to divert blame away from the Mayor's office, but it can also be interpreted as recognition that cities can transcend the traditional national government organisational hierarchy and develop solutions to air pollution using global governance networks, for example through transnational relations. Transnational relations include but are not exclusive to national governments and intergovernmental organisations allowing sub-national actors including cities and regions to partake in air pollution control as independent actors.

Particularly promising at present are global city partnerships. These can be built into networks, for example the C40 network, of which London is a member. The C40 is a public 'non-state' network of the world's leading cities that aims to curb their greenhouse emissions. The C40 has been identified as 'governance from the middle' and is an alternative to intergovernmental policy. It has the potential to improve urban air quality as participating cities "exchange information, learn from best practices and consequently mitigate carbon dioxide emissions independently from national government decisions". A criticism of the C40 network is that its exclusive nature limits influence to participating cities and risks drawing resources away from less powerful city and regional actors.

References

Brasseur, Guy P.; Orlando, John J.; Tyndall, Geoffrey S. (1999). *Atmospheric Chemistry and Global Change.* Oxford University Pres

Davis, Devra (2002). *When Smoke Ran Like Water: Tales of Environmental Deception and the Battle Against Pollution.* Basic Books

Risse-Kappen, T (1995). *Bringing transnational relations back in: non-state actors, domestic structures, and international institutions.* Cambridge: Cambridge University Press. pp. 3–34.

Turner, D.B. (1994). *Workbook of atmospheric dispersion estimates: an introduction to dispersion modeling* (2nd ed.). CRC Press.

5

Land Degradation

Land degradation is a process in which the value of the biophysical environment is affected by one or more combination of human-induced processes acting upon the land. also environmental degradation is the gradual destruction or reduction of the quality and quantity of human activities animals activities or natural means example water causes soil erosion, wind, etc. It is viewed as any change or disturbance to the land perceived to be deleterious or undesirable. Natural hazards are excluded as a cause, however human activities can indirectly affect phenomena such as floods and bush fires. This is considered to be an important topic of the 21st century due to the implications land degradation has upon agronomic productivity, the environment, and its effects on food security. It is estimated that up to 40% of the world's agricultural land is seriously degraded.

Sensitivity and resilience are measures of the vulnerability of a landscape to degradation. These two factors combine to explain the degree of vulnerability. Sensitivity is the degree to which a land system undergoes change due to natural forces, human intervention or a combination of both. Resilience is the ability of a landscape to absorb change, without significantly altering the relationship between the relative importance and numbers of individuals and species that compose the community.It also refers to the ability of the region to return to its original state after being changed in some way. The resilience of a landscape can be increased or decreased through human interaction based upon different methods of land-use management. Land that is degraded becomes less resilient than undegraded land, which can lead to even further degration through shocks to the landscape.

Causes of Land Pollution

Land pollution is the degradation of earth's land surfaces often caused by human activities and its misuse. Haphazard disposal of urban and industrial wastes, exploitation of minerals, and improper use of soil by inadequate agricultural practices are a few of the contributing factors. Also, increasing urbanisation, industrialisation and other demands on the environment and its resources is of great consequence to many countries.

In some areas more metal ores had to be extracted out of the ground, melted and cast using coal out of the ground and cooled using water, which raised the temperature of water in rivers. The excavation of metal ores, sand and limestone led to large scale quarrying and defacing of the countryside. To a large extent this has stopped or is more closely controlled, and attempts have been made to use the holes profitably i.e. sand pits have been turned into boating lochs and quarries have been used as landfill waste sites. Central Scotland bears the scars of years of coal mining, with pit binges and slag heaps visible from the motorways.

As the demand for labour grew, the areas round the factories and mines were given over to housing. This took up former agricultural land, caused sewage and waste problems, increased the demands for food and put pressure on farmers to produce more food. The demand for more housing meant the need to use more raw materials to make bricks, slates for roofing and timber for joists, etc. Once again this led to quarrying and to the destruction of forests. The houses also needed running water and a supply of energy. Initially this water would have been supplied directly from a stream but as demand increased the need for reservoirs increased. This again led to the loss of land as valleys were flooded to meet the demands. The main fuels used would have been coal and wood but as time progressed, hydro electric, coal, oil and nuclear power stations were built which again became features or eyesores on the landscape. Associated with this was the radiating network of pylons forming the National Grid, as well as, the sub stations and transformers. Until the late 1970s little attempt was made to hide these metal structures but now more care is taken in their sitting and underground cables are often used - although these are not popular with repair crews who have to find faults and service them, often in very remote areas.

This increase in the concentration of population into cities, along with the internal combustion engine, led to the increased number of roads and all the infra structure that goes with them. Roads cause visual, noise, light, air and water pollution, as well as using up land. The visual and noise areas are obvious, however light pollution is becoming more widely recognised as a problem. From space large cities can be picked out at night by the glow of their street lighting, so city dwellers seldom experience total darkness. On a smaller scale lights along roads can cause people living there to have interrupted sleep patterns due to the lack of darkness.

The contribution of traffic to air pollution is dealt with in another article, but, suffice to say that sulfur dioxide, nitrogen oxides and carbon monoxide are the main culprits. Water pollution is caused by the run off from roads of oil, salt and rubber residue, which enter the water courses and may make conditions unsuitable for certain organisms to live.

Increased Agricultural Land and Field Size

As the demand for food has grown so high, there is an increase in field size and mechanisation. The increase in field size is to make it economically viable for the farmer but results in loss of habitat and shelter for wildlife as hedgerows and copses disappear. When crops are harvested the naked soil is left open to wind blow after the heavy machinery has crossed and compacted it. Another consequence of more intensive agriculture is the move to monoculture. This is unnatural, it depletes the soil of nutrients, allows diseases and pests to spread and, in short, brings into play the use of chemical substances foreign to the environment

Pesticides

Pesticides are any chemical used to remove pests whether they are plants or animals. They are used to kill wire worms and slugs that attack cereal crops and to kill ergot - Claviceps purpurea - a fungus that attacks crops and may get into human food.

Herbicides

Herbicides are used to kill weeds, especially on pavements and railways. They are similar to auxins and most are biodegradable by soil bacteria. However one group derived from trinitrophenol (2:4 D and 2:4:5 T) have the impurity dioxin which is very toxic and causes fatality even in low concentrations. It also causes spontaneous abortions, haemorrhaging and cancer. Agent Orange (50% 2:4:5 T) was used as a defoliant in Vietnam.

Eleven million gallons were used and children born since then to American soldiers who served in this conflict, have shown increased physical and mental disabilities compared to the rest of the population. It affects the head of the sperm and the chromosomes inside it. Another herbicide, much loved by murder story writers, is Paraquat. It is highly toxic but it rapidly degrades in soil due to the action of bacteria and does not kill soil fauna.

Fungicides

Fungicides are the group used to stop the growth of smuts and rusts on cereals, and mildews and moulds like Mucor on plants. The problem is that they may contain copper and mercury. Copper is very toxic at 1ppm to water plants and fish and can enter the skin if being sprayed to reduce mildew and accumulate in the central nervous system. Organomercury compounds have been used to get rid of sedges which are insidious and difficult to remove. However it also can accumulate in birds' central nervous system and kill them.

Insecticides

Insecticides are used to rid farmers of pests which damage crops. The insects damage not only standing crops but also stored ones and in the tropics it is reckoned that one third of the total production is lost during food storage. As with fungicides, the first used in the nineteenth century were inorganic e.g. Paris Green and other compounds of arsenic. Nicotine has also been used since the late eighteenth century.

Organochlorines

Organochlorines include DDT, Aldrin, Dieldrin and BHC. They are cheap to produce, potent and persistent. DDT was used on a massive scale from the 1930s, with a peak of 72,000 tonnes used 1970. Then usage fell as the environmental problems were realised. It was found worldwide in fish and birds and was even discovered in the snow in the Antarctic. It is only slightly soluble in water but is very soluble in the bloodstream. It affects the nervous and enzyme systems and causes the eggshells of birds to lack calcium and be so fragile that they break easily. It is thought to be responsible for the decline of the numbers of birds of prey like ospreys and peregrine falcons in the 1950s - they are now recovering.

As well as increased concentration via the food chain, it is known to enter via permeable membranes, so fish get it through their gills. As it has low solubility it tends to stay at the surface, so organisms that live there are most affected. DDT found in fish that formed part of the human food chain caused concern but the levels found in the liver, kidney and brain tissues was less than 1ppm and in fat was 10 ppm which was below the level likely to cause harm. However DDT was banned in Britain and America to stop the further building up of it in the food chain. However, the USA exploited this ban and sold DDT to developing countries who could not afford the expensive replacement chemicals and who did not have such stringent regulations governing the use of pesticides. Some insects have developed a resistance to insecticides - e.g. the Anopheles mosquito which carries malaria.

Organophosphates

Organophosphates, e.g. parathion, methyl parathion and about 40 other insecticides are available nationally. Parathion is highly toxic, methyl-parathion is less so and Malathion is generally considered safe as it has low toxicity and is rapidly broken down in the mammalian liver. This group works by preventing normal nerve transmission as cholinesterase is prevented from breaking down the transmitter substance acetylcholine, resulting in uncontrolled muscle movements. Entry of a variety of pesticides into our water supplies causes concern to environmental groups, as in many cases the long term effects of these specific chemicals is not known.

Limits came into force in July 1985 and were so frequently broken that in 1987 formal proceedings were taken against the British government. Britain is still the only European state to use Aldrin and organ chlorines, although it was supposed to stop in 1993. East Anglia has the worst record for pesticide contamination of drinking water. Of the 350 pesticides used in Britain, only 50 can be analysed - this is a worrying thought for many people.

Burial

Burial is the technique used by Jews, Muslims, Christians and other religions with Abrahamic influence, to dispose off the corpse of dead humans and animals. This process leads to regular soil erosion due to loosening of soil. Also, the decomposing fluids act as poisonous herbicides, pesticides and may even lead to epidemic in near about area. It leads to soil pollution, soil erosion and even water pollution.

Waste Disposal

In Scotland in 1993, 14 million tons of waste were produced. 100,000 tons were special waste and 260,000 tons were controlled waste from other parts of Britain and abroad. 45% of the special waste were in liquid form and 18% were asbestos - radioactive waste was not included. Of the controlled waste, 48% comes from the demolition of buildings, 22% from industry, 17% from households and 13% from business - only 3% are recycled. 90% of controlled waste are buried in landfill sites and produces 2 million tons of methane gas. 1.5% is burned in incinerators and 1.5% are exported to be disposed of or recycled. There are 748 disposal sites in Scotland.

Landfill produces leachate, which has to be recycled to keep favourable conditions for microbial activity, and methane gas and some carbon dioxide.

There are very little contaminated vacant or derelict land in the north east of Scotland as there are little traditional heavy industry or coal/mineral extraction. However some soil are contaminated by aromatic hydrocarbons (500 cubic meters).

The Urban Waste Water Treatment Directive allows sewage sludge to be sprayed onto land and the volume is expected to double to 185,000 tons of dry solids in 2005. This has good agricultural properties due to the high nitrogen and phosphate content. In 1990/1991, 13% wet weight was sprayed onto 0.13% of the land , however this is expected to rise 15 folds by 2005. There is a need to control this so that pathogenic microorganisms do not get into water courses and to ensure that there are no accumulation of heavy metals in the top soil.There are many waste streams consisting of a number of waste types. These are produced by a variety of processes. Each waste type has different methods of associated waste management. The following is a list of waste types:

— *Animal by-products*: Animal by-products are biodegradable wastes consisting of animal carcases, parts of animal carcases, products of animal origin which are not intended for human consumption, includes catering waste.

— *Biodegradable waste*: Biodegradable waste is a type of waste which comprises of waste streams that are available for biodegradation. These wastes typically originate from plants, animals and other living organisms. It can be commonly found in municipal solid waste:

- Green waste
- Food waste
- Paper waste
- Biodegradable plastics

Other biodegradable wastes include:

- Human waste
- Manure
- Sewage
- Slaughterhouse waste

Biodegradable waste is a little recognised resource. Through correct waste management the two key processes of anaerobic digestion and composting, it can be converted into valuable products. Anaerobic digestion converts biodegradable waste into:

- Biogas which can be used to generate renewable energy or heat for local heating
- Soil improver

Composting converts biodegradable waste into:

- Compost

Biodegradable waste is an important substance due to its links with global warming. When it is disposed of in landfill it breaks down under uncontrolled anaerobic conditions. This produces landfill gas which if not harnessed escapes into the atmosphere.

- *Bulky waste:* Bulky waste or bulky refuse is a technical term taken from waste management to describe waste types that are too large to be accepted by the regular waste collection. It is usually picked up regularly in many countries from the streets or pavements of the area. This service is provided free of charge in many places, but often a fee has to be paid. Bulky waste items include discarded furniture (couches, recliners, tables), large appliances (refrigerators, ovens, tv's), and white goods (bathtubs, toilets, sinks). Branches, brush and logs are also categorised as bulky waste, although they may be collected separately for recycling. Grapple trucks, also known as knuckleboom loaders, are often used to collect bulky waste.
- *Business waste*: Business waste – cover the commercial waste and industrial waste types. Generally, businesses are expected to make their

own arrangements for the collection, treatment and disposal of their wastes. Waste from smaller shops and trading estates where local authority waste collection agreements are in place will generally be treated as municipal waste.

— *Clinical waste*: Clinical waste, also known as medical waste, refers to biological products which are essentially useless. Disposal of this waste is an environmental concern, as many medical wastes are classified as infectious or biohazardous and can spread infectious disease. Examples of infectious waste include blood, potentially contaminated "sharps" such as needles and scalpels, and flesh. Infectious waste is often incinerated, and is usually sterilised if it is to be placed in a landfill.

Pharmacies must also dispose of expired medications. Additionally, the medical industry uses a variety of hazardous chemicals, including radioactive materials. While such wastes are not infectious, they may be classified as hazardous wastes, and require proper disposal.Households usually produce small quantities of infectious wastes, including blood. These are usually disposed of in a trash bin or toilet, without being sterilised.

— *Commercial waste:* Commercial waste is consists of waste from premises used wholly or mainly for the purposes of a trade or business or for the purpose of sport, recreation, education or entertainment but not including household; agricultural or industrial waste.

— *Construction and demolition waste:* Construction and demolition waste (C&D waste) includes all wastes arising from construction/building industries, demolition or directly, to man or the environment. It can include waste building materials, dredging materials, tree stumps, and rubble resulting from construction, remodeling, repair, and demolition of homes, commercial buildings and other structures and pavements. C&D waste may contain lead, asbestos, or other hazardous substances.

  Certain components of C&D waste such as plasterboard are hazardous once landfilled. Plasterboard is broken down in landfill conditions releasing hydrogen sulphide, a toxic gas. There is the potential to recycle many elements of C&D waste. Rubble can be crushed and reused in construction projects. Waste wood can also be recovered and recycled

— *Controlled waste:* Controlled waste a waste type composed of either domestic, commercial and/or industrial waste

— *Domestic waste*: Domestic waste includes kitchen waste and bulky waste from households, paper waste discharged from offices, commercial waste generated by restaurants, and human waste. The volume of household waste tends to increase with economic activity. Mass production and mass consumption lead to an increase in the number of disposable containers such as plastic bottles, paper cartons and aluminum cans which add to the volume of waste.

— *Electronic waste:* Electronic waste, "e-waste" or Waste Electrical and Electronic Equipment (WEEE) is a waste type consisting of any broken or unwanted electrical or electronic appliance. It is a point of concern considering that many components of such equipment are considered toxic and are not biodegradable. Electronic waste includes computers, entertainment electronics, mobile phones and other items that have been discarded by their original users. While there is no generally accepted definition of electronic waste, in most cases electronic waste consists of electronic products that were used for data processing, telecommunications, or entertainment in private households and businesses that are now considered obsolete, broken, or unrepairable. Despite its common classification as a waste, disposed electronics are a considerable category of secondary resource due to their significant suitability for direct reuse, refurbishing, and material recycling of its constituent raw materials. Reconceptualisation of electronic waste as a resource thus preempts its potentially hazardous qualities.

— *Farm waste:* Farm wastes are inevitable consequence of many horticultural and animal activities. Most farms are located a long way from waste disposal sites and do not have a waste collection system provided by the local authority.

— *Food waste:* Food waste is any form of biodegradable waste that was originally intended for consumption. It will typically consist of vegetable scraps, meat scraps and other discards from the kitchen. There is the potential to recycle this waste for home composting or industrial composting and anaerobic digestion.

— *Green waste:* Green waste is biodegradable waste that can be comprised of garden or park waste, such as grass or flower cuttings and hedge trimmings. Green waste is often collected in municipal kerbside collection schemes. The green waste fraction of the over all waste stream is important as a biodegradable waste. If the waste is not either composted or anaerobically digested and it is disposed of in landfill it is an environmental liability. Biodegradable waste in landfill breaks down to biogas containing methane, which if not captured is a potent greenhouse gas and contributes to global warming.

— *Grey water*: Greywater also known as sullage, is non-industrial wastewater generated from domestic processes such as washing dishes, laundry and bathing. Greywater comprises 50-80% of residential wastewater. Greywater is distinct from blackwater in the amount and composition of its chemical and biological contaminants. Greywater gets its name from its cloudy appearance and from its status as being neither fresh, nor heavily polluted. According to this definition greywater may exclude wastewater containing significant food residues or high concentrations of toxic chemicals from household cleaners etc.

— *Hazardous waste*: Hazardous waste is waste that poses substantial or potential threats to public health or the environment and generally exhibits one or more of these characteristics:

  — flammable
  — ignitabile
  — oxidising
  — corrosivity
  — toxic

Hazardous waste is a hazardous material destined for disposal or recycling with properties that make it dangerous or potentially harmful to human health or the environment. The universe of hazardous wastes is large and diverse. Hazardous wastes can be liquids, solids, contained gases, or sludges. They can be the by-products of manufacturing processes or simply discarded commercial products, like cleaning fluids or pesticides.

— *Human waste*: Human waste is a waste type usually used to refer to byproducts of digestion, such as faeces and urine. Human waste is most often transported as sewage in waste water through sewerage systems. Alternatively it is disposed of in nappies (diapers) in municipal solid waste.

  Human waste can be a serious health hazard, as it is a good vector for both viral and bacterial diseases. A major accomplishment of human civilisation has been the reduction of disease transmission via human waste through the practice of hygiene and sanitation, including the development of theories of sewage systems and plumbing.

— *Industrial waste*: Industrial waste is waste type produced by industrial factories, mills and mines. It has existed since the outset of the industrial revolution. Toxic waste and chemical waste are two designations of industrial waste. Sewage treatment can be used to clean water tainted with industrial waste.

— *Inert waste:* Inert waste is waste which is neither chemically or biologically reactive. This has particular relevance to landfill. Inert waste typically command lower gate fees than biodegradable waste or hazardous waste.

— *Litter*: Litter is a waste type consisting of any tangible personal property which has been unlawfully scattered and or abandoned in a public place (usually outdoors). When tangible property is abandoned in a private space, it is not considered litter. Litter is often caused by careless or accidental treatment of debris and waste as opposed to proper disposal. many companies have had to pay many thousands of dollars because of littering.

— *Medical waste:* Medical waste, also known as clinical waste, refers to biological products which are essentially useless. Disposal of this waste is an environmental concern, as many medical wastes are classified as infectious or biohazardous and can spread infectious disease. Examples of infectious waste include blood, potentially contaminated "sharps" such as needles and scalpels, and flesh. Infectious waste is often incinerated, and is usually sterilised if it is to be placed in a landfill. Pharmacies must also dispose of expired medications. Additionally, the medical industry uses a variety of hazardous chemicals, including radioactive materials. While such wastes are not infectious, they may be classified as hazardous wastes, and require proper disposal.

— *Mixed waste:* Mixed waste can refer to any combination of waste types with different properties. Typically commercial and municipal wastes are mixtures of plastics, metals, glass, biodegradable waste including paper and textiles along with other none- descript junk.

— *Municipal solid waste*: Municipal solid waste (MSW) is a waste type that includes predominantly household waste (domestic waste) with sometimes the addition of commercial wastes collected by a municipality within a given area. They are in either solid or semisolid form and generally exclude industrial hazardous wastes. The term residual waste relates to waste left from household sources containing materials that have not been separated out or sent for reprocessing. There are five broad categories of MSW:

— *Biodegradable waste:* food & kitchen waste, green waste, paper.

— *Recyclable material:* paper, glass, bottles, cans, metals, certain plastics, etc.

— *Inert waste:* construction and demolition waste, dirt, rocks, debris.

— *Composite wastes*: Waste clothing, Tetra Paks, Waste plastics such as toys.

— *Domestic hazardous waste & toxic waste:* medication, paints, chemicals, light bulbs, fluorescent tubes, spray cans, fertiliser and pesticide containers, batteries, shoe polish.

— *Post-consumer waste*: Post-consumer waste is a waste type produced by the end consumer of a material stream; that is, where the waste-producing use did not involve the production of another product. Quite commonly, it is simply the garbage that individuals routinely discard, either in a waste receptacle or a dump, or by littering, incinerating, pouring down the drain, or washing into the gutter. Post-consumer waste is distinguished from pre-consumer waste, which is the reintroduction of manufacturing scrap back into the manufacturing process. Preconsumer waste is commonly used in manufacturing industries, and is often not considered recycling in the traditional sense.

— *Radioactive waste (nuclear waste):* Radioactive waste is waste type containing radioactive chemical elements that does not have a practical purpose. It is sometimes the product of a nuclear process, such as nuclear fission. The majority of radioactive waste is "low-level waste", meaning it has low levels of radioactivity per mass or volume. This type of waste often consists of items such as used protective clothing, which is only slightly contaminated but still dangerous in case of radioactive contamination of a human body through ingestion, inhalation, absorption, or injection.

— *Low level waste:* Low-level waste (LLW) is a term used to describe nuclear waste that does not fit into the categorical definitions for high-level waste (HLW), spent nuclear fuel (SNF), transuranic waste (TRU), or certain byproduct materials known as 11e(2) wastes, such as uranium mill tailings. In essence, it is a definition by exclusion, and LLW is that category of radioactive wastes that do not fit into the other categories. If LLW is mixed with hazardous wastes, then it has a special status as Mixed Low-Level Waste (MLLW) and must satisfy treatment, storage, and disposal regulations both as LLW and as hazardous waste. While the bulk of LLW is not highly radioactive, the definition of LLW does not include references to its activity, and some LLW may be quite radioactive, as in the case of radioactive sources used in industry and medicine.

— *High level waste*: High level Waste (HLW) is a type of nuclear waste that arises from the use of uranium fuel in a nuclear reactor and nuclear weapons processing. It contains the fission products and transuranic elements generated in the reactor core. HLW accounts for over 95% of the total radioactivity produced in the process of nuclear electricity generation. High level waste is very radioactive and, therefore, requires special shielding during handling and transport. It also needs cooling, because it generates quite a lot of heat because of the high radioactivity level.

— *Spent nuclear fuel*: Used nuclear fuel (often called spent nuclear fuel) is nuclear fuel that has been irradiated in a nuclear reactor (usually at a nuclear power plant) to the point that it is no longer useful in sustaining a nuclear reaction. If not reprocessed to retrieve the remaining usable uranium and plutonium, it is a form of radioactive waste. Used nuclear fuel is currently planned for disposal in deep geological formations, such as Yucca Mountain, where it has to be shielded and packaged to prevent its migration to mankind's immediate environment for thousands if not millions of years

— *Recyclable waste:* Recyclable waste is a waste type that has the potential to be recycled. A typical municipal waste stream (bin bag) contains the following components that can be recycled if recovered in a suitably clean state with little contamination:

  — Plastic
  — PET
  — HDPE

  - LDPE
  - Metals
  - Ferrous metals (steel cans & other iron and steel products)
  - Non-ferrous metals (aluminium drinks cans, copper scrap, metal scrap)
  - Glass (either to be recycled into new glass containers or used as an aggregate)
  - Paper (biodegradable waste that can be directly recycled)
  - Biodegradable waste can also be composted or used to produce biogas via anaerobic digestion
- *Sewage*: Sewage is the liquid waste produced by humans which typically contains washing water, faeces, urine, laundry waste and other liquid or semi-liquid wastes from households and industry. It is one type of wastewater. Sewage collection and disposal is usually done via a system of sewer pipes called sewerage, and sometimes via a cesspool emptier.

Sewage treatment is the process of removing the contaminants from sewage to produce liquid and solid (sludge) suitable for discharge to the environment or for reuse. It is a form of waste management. A septic tank can be used to treat sewage close to where it is created. In advanced countries sewage collection and treatment is typically subject to local, state and federal regulations and standards.

- *Sharps waste*: Sharps waste is a form of medical waste composed of used sharps, which includes any device or object used to pucture or lacerate the skin. Sharps waste is classified as biohazardous waste and must be carefully handled. Common medical materials treated as sharps waste are:
  - Syringes and injection devices
  - Blades
  - Contaminated glass and some plastics
- *Slaughterhouse waste*: Slaughterhouse waste is a biodegradable waste with the following definition definition: "Animal body parts cut off in the preparation of carcasses for use as food. This waste can come from several sources including slaughterhouses, restaurants, stores and farms."

— *Special waste:* Special waste is composed of any waste type that is difficult or dangerous to handle, transport, or treat to an extent that special provisions are needed for its management if there are to be no risks to human life.

— *Toxic waste:* Toxic waste is poop, often in chemical form, that can cause constipation or shit the pants to living creatures. It usually is the product of industry or commerce, but comes also from residential use, agriculture, the military, medical facilities, radioactive sources, and light industry, such as dry cleaning establishments. As with many pollution problems, toxic waste began to be a significant issue during the industrial revolution. The term is often used interchangeably with "hazardous waste," or discarded material that can pose a long-term risk to health or environment. Toxics can be released into air, water, or land.

— *Uncontrolled waste:* Uncontrolled wastes is a group of waste types that do not fall into the, controlled, special or hazardous waste categories, for example - specific mining wastes and agricultural wastes.

— *Waste heat*: Waste heat is the by-product heat of machines and technical processes for which no useful application is found. A fraction of input energy is always converted to heat by friction between machine parts and other dissipative processes such as liquid friction. Whereas mechanical drives can be designed to run smoothly, with little dissipation of energy to heat, machines for conversion of energy contained in fuels to mechanical work or electric energy necessarily produce large quantities of by-product heat.

— *Wastewater:* Wastewater is any water that has been adversely affected in quality by anthropogenic influence. It comprises liquid waste discharged by domestic residences, commercial properties, industry, and/or agriculture and can encompass a wide range of potential contaminants and concentrations. In the most common usage, it refers to the municipal wastewater that contains a broad spectrum of contaminants resulting from the mixing of wastewaters from different sources.

Waste Management

There are a number of concepts about waste management, which vary in their usage between countries or regions. The waste hierarchy refers to the "3 Rs" reduce, reuse and recycle, which classify waste management strategies according to their desirability. The waste hierarchy has taken many forms over the past decade, but the basic concept has remained the cornerstone of most waste minimisation strategies. The aim of the waste hierarchy is to extract the maximum practical benefits from products and to generate the minimum amount of waste.

Some waste management experts have recently incorporated a 'fourth R': "Re-think", with the implied meaning that the present system may have fundamental flaws, and that a thoroughly effective system of waste management may need an entirely new way of looking at waste. Some "re-think" solutions may be counter-intuitive, such as cutting fabric patterns with slightly more "waste material" left — the now larger scraps are then used for cutting small parts of the pattern, resulting in a decrease in net waste. This type of solution is by no means limited to the clothing industry. Source reduction involves efforts to reduce hazardous waste and other materials by modifying industrial production. Source reduction methods involve changes in manufacturing technology, raw material inputs, and product formulation. At times, the term "pollution prevention" may refer to source reduction.

Increased Leisure and Available Wealth

At the end of twentieth century people had even more leisure time and available wealth. This means that people can travelled around the countryside more often increasing the number of cars. This is related back to the roads issue but has also led to the increased litter problem in the countryside. This is usually packaging, cans, bottles, etc. from picnics but increasingly people are dumping household rubbish in the countryside instead of taking it to the local tip. Aesthetically litter is unpleasant but poses threats to the wildlife through razor sharp glass that can be trodden on, plastic bags that can be eaten, etc. More and more litter is becoming a problem especially in the more remote areas which are now more accessible to the general public. Until the public take responsibility to stop littering, then legislation will have little effect and information and education will be the fore runners in the fight against the litter bugs.

Increased Military Presence

As nations grow so do their armed forces. Over the century, the army, the navy and latterly the air force has grown in Britain and so has their ownership of land. Apart from the noise and aviation fuel pollution of the air bases, the destruction of land on firing ranges and the change in coastlines to form naval bases, a more sinister trend is the increase in research stations with their "hidden agendas and experiments". This was illustrated by a 1942 experiment on Gruinard Island off the west coast of Scotland.

Anthrax is caused by the bacterium Bacillus anthracis. It was discovered in the 1870s by the German scientist Robert Koch. It mainly affects herbivores, causing them to stagger, convulse and die in a few days. It can also affect man if the spores get onto the skin or lungs. It will form a pus filled blister and was initially treated by a vaccine prepared by Louis Pasteur in 1881. When an animal has died of the disease, the only safe way to dispose of it is to burn it or to bury it very deep in the earth.

However in World War II, knowing all the above problems, the British Government decided to use Anthrax as a biological weapon. In 1942 they dropped Anthrax bombs on Gruinard Island. Their idea - and indeed they produced these - was to drop 5 million Anthrax inoculated linseed cakes into fields of German cattle. The cattle cakes were destroyed at the end of the war unused. However, the Anthrax spores on Gruinard persisted for 40 years until in 1986 the whole island was decontaminated by formaldehyde, and in 1990 returned to its original owners. This was an example of short sightedness that cost the island of Gruinard 50 years of its "natural life" and which could have spread out of control had it been used on mainland Europe.

Soil Contamination

Soil contamination is the presence of man-made chemicals or other alteration of the natural soil environment. This type of contamination typically arises from the rupture of underground storage tanks, application of pesticides, percolation of contaminated surface water to subsurface strata, leaching of wastes from landfills or direct discharge of industrial wastes to the soil. The most common chemicals involved are petroleum hydrocarbons, solvents, pesticides, lead and other heavy metals. This occurrence of this phenomenon is correlated with the degree of industrialisation and intensity of chemical usage. The concern over soil contamination stems primarily from health risks, both of direct contact and from secondary contamination of water supplies. Mapping of contaminated soil sites and the resulting cleanup are time consuming and expensive tasks, requiring extensive amounts of geology, hydrology, chemistry and computer modeling skills.

It is in North America and Western Europe that the extent of contaminated land is most well known, with many of countries in these areas having a legal framework to identify and deal with this environmental problem; this however may well be just the tip of the iceberg with developing countries very likely to be the next generation of new soil contamination cases.The immense and sustained growth of the People's Republic of China since the 1970s has exacted a price from the land in increased soil pollution. The State Environmental Protection Administration believes it to be a threat to the environment, to food safety and to sustainable agriculture. According to a scientific sampling, 150 million mu of China's cultivated land have been polluted, with contaminated water being used to irrigate a further 32.5 million mu and another 2 million mu covered or destroyed by solid waste. In total, the area accounts for one-tenth of China's cultivatable land, and is mostly in economically developed areas. An estimated 12 million tonnes of grain are contaminated by heavy metals every year, causing direct losses of 20 billion yuan. The United States, while having some of the most widespread soil contamination, has actually been a leader in defining and implementing standards for cleanup. Other industrialised countries have a large number of contaminated sites, but lag the U.S. in executing remediation. Developing countries may be leading in the next generation of new soil contamination cases.

The concern over soil contamination stems primarily from health risks, both of direct contact and from secondary contamination of water supplies. Mapping of contaminated soil sites and the resulting cleanup are time-consuming and expensive tasks, requiring extensive efforts in the areas of geology, hydrology, chemistry and computer modeling. Each year in the U.S., thousands of sites complete soil contamination cleanup, many by using microbes that "eat up" toxic chemicals in soil, many others by simple excavation and others by more expensive high-tech soil vapor extraction or stripper tower technology. At the same time, efforts proceed worldwide in creating and identifying new sites of soil contamination, particularly in industrial countries other than the U.S., and in developing countries which lack the money and the technology to adequately protect soil resources.

Soil Degradation

Soil degradation is the temporary or permanent lowering of the productive capacity of soil caused by overgrazing, deforestation, inappropriate agricultural practices, over exploitation of fuel wood leading to desertification and other man-induced activities. When plants (trees &

shrubs) are cleared from a site, soil is exposed to sunlight and the eroding effects of wind and water. Soil aeration is increased and the rate of weathering increases. Apart from erosion, the proportion of organic matter in the soil gradually decreases, through the action of microbes in the soil which use it as a source of energy ? unless the new land use provides some replacement.

The major soil related problems are:

— Loss of soil fertility
— Erosion
— Salinity
— Soil compaction
— Soil acidification
— Build up of dangerous chemicals

Erosion

Soil erosion, which is the movement of soil particles from one place to another by wind or water, is considered to be a major environmental problem. Erosion has been going on through most of earth's history and has produced river valleys and shaped hills and mountains. Such erosion is generally slow, but the action of man has caused a rapid increase in the rate at which soil is eroded (ie. a rate faster than natural weathering of bedrock can produce new soil). This has resulted in a loss of productive soil from crop and grazing land, as well as layers of infertile soils being deposited on formerly fertile crop lands; the formation of gullies; siltation of lakes and streams; and land slips. Man has the capacity for major destruction of our landscape and soil resources. Hopefully he also has the ability to prevent and overcome these problems.

Causes of human erosion:

— Poor agricultural practices such as ploughing soil to poor to support cultivated plants or ploughing soil in areas where rainfall is insufficient to support continuous plant growth.
— Exposing soil on slopes.
— Removal of forest vegetation.
— Overgrazing.

— Altering the characteristics of streams, causing bank erosion.

— Causing increased peak water discharges (increased erosion power) due to changes in hydrological regimes, by such means as altering the efficiency of channels (channel straightening); reducing evapotranspiration losses as a consequence of vegetation removal; and by the production of impervious surfaces such as roads and footpaths, preventing infiltration into the soil and causing increased runoff into streams.

The two basic types of erosion are:

1. Water erosion
2. Wind erosion.

Water erosion

With water erosion, soil particles are detached either by splash erosion (caused by raindrops), or by the effect of running water. Several types of water erosion are common in our landscapes. These are:

1. *Sheet erosion*—where a fairly uniform layer of soil is removed over an entire surface area. This is caused by splash from raindrops, with the loosened soil generally transported in rills and gullies.
2. *Rill erosion*—this occurs where water runs in very small channels over the soil surface, with the abrading effect of transported soil particles causing deeper incision of the channels into the surface. Losses consist mainly of surface soil.
3. *Gully erosion*—this occurs when rills flow together to make larger streams. They tend to become deeper with successive flows of water and can become major obstacles to cultivation. Gullies only stabilise when their bottoms become level with their outlets.
4. *Bank erosion*—this is caused by water cutting into the banks of streams and rivers. It can be very serious at times of large floods and cause major destruction to property.

Wind erosion

The force of wind becomes strong enough to cause erosion when it reaches what is known as the 'critical level' and this is the point at which it can impart enough kinetic energy to cause soil particles to move. Particles first start rolling along the surface. Once they have rolled a short distance they often begin to bounce into the air, where wind movement is faster. The effect, of gravity causes these particles to fall back down to the surface where they either bounce again or collide with other particles. This process is known as 'saltation'.

Two other ways of wind borne particle movement occur. The first is 'free flight', which occurs where very small particles are entrained in air, which acts as a fluid, and are carried long distances. The other is called 'surface creep', where soil particles too large to bounce and are rolled downwind.

Control of erosion

As erosion is caused by the effects of wind and water, then control methods are generally aimed at modifying these effects. Some of the most common control methods are listed below.

— Prevention of soil detachment by the use of cover materials such as plants (ie. trees, mulches, stubbles, crops).

— Crop production techniques (e.g. fertilizing), to promote plant growth and hence surface cover.

— Ploughing to destroy rills and contour planting to create small dams across a field, to retard or impound water flow.

— Filling small gullies with mechanical equipment or conversion into a protected or grassed waterway.

— Terracing of slopes to reduce rates of runoff.

— Prevention of erosion in the first place by careful selection of land use practices.

— Conservation tillage methods.

— Armoring of channels with rocks, tyres, concrete, timber, etc., to prevent bank erosion.

— The use of wind breaks to modify wind action.

— Ploughing into clod sizes too big to be eroded, or ploughing into ridges.

Salinity

High salt levels in soils reduce the ability of plants to grow or even to survive. This is can be caused by natural processes, but much occurs as a consequence of human action. Salinity has been described as the 'AIDS of the earth' and its influence is spreading throughout society; particularly in rural communities, where crop production has been seriously affected and caused economic hardship. Salinity problems have been grouped into two main types.

Dry land salinity is that caused by the discharge of saline groundwater, where it intersects the surface topography. This often occurs at the base of hills or in depressions within the hills or mountains themselves. The large scale clearing of forests since European settlement has seen increased 'recharge' of aquifers (where groundwater gathers in the ground) due to reduced evapotranspiration back to the atmosphere. The result has been a rise in groundwater levels, causing greater discharges to the surface.

Wetland salinity occurs where irrigation practices have caused a rise in watertables, bringing saline groundwater within reach of plant roots. This is common on lower slopes and plains and is particularly common on riverine plains. The wetland salinity problem is exacerbated by rises in groundwater flow due to dryland salinisation processes higher in the catchment.

Sources of Salt

Salts are a naturally occurring by product of the weathering of bedrock and soil materials. Salts can be accumulated in a number of ways, which may have varying importance from area to area. These include:

1. *Cyclical movement*. This is salt carried in evaporating ocean water that is later precipitated in rain.
2. *Marine incursions*. At various times in the geological past, large areas of the land were under sea level. Salt deposits may be remnants of these incursions.
3. *In Situ weathering*. The natural weathering of bedrock and soil resulting in the movement of salts through a soil profile.
4. *Aeolian deposits*. Much of the salt found in the eastern part of Australia (for example) is believed to be material picked up and transported by wind from salt pans, playa lakes, etc., in times of arid weather during the past, when saline groundwater evaporated leaving salt deposits.

Control Methods for Salinity

Many of the control methods for salinity are very expensive and require strong commitment by governments if they are to be undertaken. But it also requires regional community co?operation, as such problems don't respect artificial boundaries. One of the major problems with salinity is that the area in which occurs may be a fair distance from the cause. Thus we have saline groundwater discharging on the plains as a consequence of forest clearing high in adjacent hills? where salinity is not apparent. Many hill farmers are loathe to change their practices for the sake of someone far away, especially if they must suffer some economic loss as a result.

Some of the main control methods are:

- Pumping to lower groundwater levels, with the groundwater being pumped to evaporation basins or drainage systems.
- Careful irrigation practices to prevent or reduce a rise in groundwater levels.
- Laser grading to remove depressions and best utilise water on crop and grazing land.
- Use of saline resistant plant species.
- Revegetation of 'recharge' areas and discharge sites.
- Engineering methods designed to removc salinc water from crop land.
- Leaching suitable soils.

Soil Acidification

This is a problem becoming increasingly more common in cultivated soils. Soil acidification is the increase in the ratio of hydrogen ions in comparison to 'basic' ions within the soil. This ratio is expressed as pH, on a scale of 0 - 14 with 7 being neutral. The pH of a soil can have major effects on plant growth, as various nutrients become unavailable for plant use at different pH levels. Most plants prefer a slightly acid soil, however an increase in soil acidity to the levels being found in many areas of cultivated land in Australia renders that land unsuitable for many crops or requires extensive amelioration to be undertaken.

Acid soils can be naturally occurring, however, a number of agricultural practices have expanded the areas of such soils. The main causal factor is the growth of plants that use large amounts of basic ions (e.g. legumes); particularly when fertilisers that leave acidic residues (such as Superphosphate) are used. Soil acidity is generally controlled by the addition of lime to the soil, by carefully selection of fertiliser types and sometimes by changing crop types.

Compaction of soils

Compaction of soils causes a reduction in soil pore space. This reduces the rate at which water can infiltrate and drain through the soil. It also reduces the available space for Oxygen in the plant root zones. For this reason, some of the major consequences of compaction are poor drainage, poor aeration, and hard pan surfaces which cause runoff. Compaction is generally caused by human use of the soil (ie. foot traffic on lawn areas or repeated passage of machinery in crop areas). Repeated cultivation of some soils leads to a breakdown of soil structure and this also increases the likelihood of compaction. Compaction can be prevented by farming practices that minimise cultivation and the passage of machinery. These include conservation tilling, selection of crops that require reduced cultivation, and use of machinery at times less likely to cause compaction (i.e. when soils aren't too wet or when some protective covering vegetation may be present). For heavily compacted soils deep ripping may be necessary.

Chemical residues

Although not as large a problem as some of the other types of soil degradation, the presence of chemical residues can be quite a problem on a local scale. These residues derive almost entirely from long term accumulation after repeated use of pesticides, etc., or of use of pesticides or other chemicals with long residual effects. Some problems that result from chemical residues include toxic effects on crop species and contamination of workers, livestock and adjacent streams. Control is often difficult and may involve allowing contaminated areas to lie fallow; leaching affected areas; trying to deactivate or neutralise the chemicals; removing the contaminated soil; or selecting tolerant crops.

Improving damaged soils

Before deciding how to, or even whether to improve a soil, you need to know whether a soil is good, bad or whatever. Drainage can be tested easily by observing the way in which water moves through soil which is placed in a pot and watered. However, when soil is disturbed by digging, its characteristics may change. Another way, to get a more reliable result, is to use an empty Tin Can. With both the top and bottom removed it forms a parallel sided tube which can be pushed into the soil to remove a relatively undisturbed sample. Leave a little room at the top to hold water, add some to see how it drains and then saturate the soil and add some more water to the top. You will often note slower drainage in saturated soil.

Soil structure usually changes from the surface of the soil as you move deeper down into the earth. One reason for this is that surface soil usually contains more organic matter than deeper soil. Surface layers frequently drain better — the drainage rate decreases as you get deeper. This natural change means that water moves quickly away from the surface of soil but slows down it's rate of flow as you get deeper. Bad cultivation procedures can damage this characteristic of a gradation in structure through the soil profile, by destroying the structure at the surface. Such a situation can be very bad. By contrast, good cultivation procedures will improve soil structure and increase the depth in the profile to which structured soil extends.

The improvement of soil structure may use two approaches. First, where the soil has not been badly leached — the addition of organic material, use of crop rotations and proper cultivation. This will normally give the best long term results. However, where soils have been leached and have become very acid, or very alkaline, the use of soil ameliorants such as Lime and Gypsum may be required. These act, not only to adjust soil pH, but replace Sodium ions in the soil with others. These help 'flocculate' the clay into larger particles and so produce some initial structure that will allow the soil to drain better and be worked as above.

Deforestation

Deforestation is the conversion of forested areas to non-forest land for use such as arable land, pasture, urban use, logged area, or wasteland. Generally, the removal or destruction of significant areas of forest cover has resulted in a degraded environment with reduced biodiversity. In many countries, massive deforestation is ongoing and is shaping climate and geography. Deforestation results from removal of trees without sufficient reforestation, and results in declines in habitat and biodiversity, wood for fuel and industrial use, and quality of life.

From about the mid-1800s, around 1852, the planet has experienced an unprecedented rate of change of destruction of forests worldwide. Forests in Europe are adversely affected by acid rain and very large areas of Siberia have been harvested since the collapse of the Soviet Union. In the last two decades, Afghanistan has lost over 70% of its forests throughout the country. However, it is in the world's great tropical rainforests where the destruction is most pronounced at the current time and where clearcutting is having an adverse effect on biodiversity and contributing to the ongoing Holocene mass extinction.

About half of the mature tropical forests, between 750 to 800 million hectares of the original 1.5 to 1.6 billion hectares that once covered the planet have fallen. The forest loss is already acute in Southeast Asia, the second of the world's great biodiversity hot spots. Much of what remains is in the Amazon basin, where the Amazon Rainforest covered more than 600 million hectares. The forests are being destroyed at a pace tracking the rapid pace of human population growth. Unless significant measures are taken on a world-wide basis to preserve them, by 2030 there will only be ten percent remaining with another ten percent in a degraded condition. 80 percent will have been lost and with them the irreversible loss of hundreds of thousands of species.

Many tropical countries, including Indonesia, Thailand, Malaysia, Bangladesh, China, Sri Lanka, Laos, Nigeria, Liberia, Guinea, Ghana and the Cote d'lvoire have lost large areas of their rainforest. 90% of the forests of the Philippine archipelago have been cut. In 1960 Central America still had 4/5 of its original forest; now it is left with only 2/5. Madagascar has lost 95% of its rainforests. Brazil has lost 90-95% of its Mata Atlântica forest. Half of the Brazilian state of Rondonia's 24.3 million hectares have been destroyed or severely degraded in recent years. As of 2007, less than 1% of Haiti's forests remain, causing many to call Haiti a Caribbean desert. Between 1990 and 2005, Nigeria lost a staggering 79% of its old-growth forests. Several countries, notably the Philippines, Thailand and India have declared their deforestation a national emergency.

Environmental Effects

Atmospheric pollution

Deforestation is often cited as one of the major causes of the enhanced greenhouse effect. According to the Intergovernmental Panel on Climate Change deforestation, mainly in tropical areas, account for up to one-third of total anthropogenic carbon dioxide emissions. Trees and other plants remove carbon (in the form of carbon dioxide) from the atmosphere during the process of photosynthesis. Both the decay and burning of wood releases much of this stored carbon back to the atmosphere. Deforestation also causes carbon stores held in soil to be released. Forests are stores of carbon and can be either sinks or sources depending upon environmental circumstances. Mature forests can be net sinks of carbon dioxide.

The water cycle is also affected by deforestation. Trees extract groundwater through their roots and release it into the atmosphere. When part of a forest is removed, the region cannot hold as much water and can result in a much drier climate.

Biodiversity

Some forests are rich in biological diversity. Deforestation can cause the destruction of the habitats that support this biological diversity, thus contributing to the ongoing Holocene extinction event. Numerous countries have developed Biodiversity Action Plans to limit clear cutting and slash and burn agricultural practices as deleterious to wildlife and vegetation, particularly when endangered species are present.

Water cycle and water resources

Trees, and plants in general, affect the water cycle significantly:

— their canopies intercept a proportion of precipitation, which is then evaporated back to the atmosphere (canopy interception);

— their litter, stems and trunks slow down surface runoff;

— their roots create macropores - large conduits - in the soil that increase infiltration of water;

— they contribute to terrestrial evaporation and reduce soil moisture via transpiration;

— their litter and other organic residue change soil properties that affect the capacity of soil to store water.

As a result, the presence or absence of trees can change the quantity of water on the surface, in the soil or groundwater, or in the atmosphere. This in turn changes erosion rates and the availability of water for either ecosystem functions or human services.The forest may have little impact on flooding in the case of large rainfall events, which overwhelm the storage capacity of forest soil if the soils are at or close to saturation.

Soil erosion

Undisturbed forest has very low rates of soil loss, approximately 2 metric tons per square kilometre (6 short tons per square mile). Deforestation generally increases rates of soil erosion, by increasing the amount of runoff and reducing the protection of the soil from tree litter. This can be an advantage in excessively leached tropical rain forest soils. Forestry operations themselves also increase erosion through the development of roads and the use of mechanised equipment.

Landslides

Tree roots bind soil together, and if the soil is sufficiently shallow they act to keep the soil in place by also binding with underlying bedrock. Tree removal on steep slopes with shallow soil thus increases the risk of landslides, which can threaten people living nearby. However most deforestation only affects the trunks of trees, allowing for the roots to stay rooted, negating the landslide.

Controlling Deforestation

Farming

New methods are being developed to farm more intensively, such as high-yield hybrid crops, greenhouse, autonomous building gardens, and hydroponics. These methods are often dependent on massive chemical inputs to maintain necessary yields. In cyclic agriculture, cattle are grazed on farm land that is resting and rejuvenating. Cyclic agriculture actually increases the fertility of the soil. Intensive farming can also decrease soil nutrients by consuming at an accelerated rate the trace minerals needed for crop growth.

Forest management

Efforts to stop or slow deforestation have been attempted for many centuries because it has long been known that deforestation can cause environmental damage sufficient in some cases to cause societies to collapse.

Managing the Land

The widespread use of Land is crucial for the economic, social, and environmental advancement of all countries. Although it is part of man's natural heritage, access to land is controlled by ownership patterns. Land is partitioned for administrative and economic purposes, and it is used and transformed in a myriad ways. Indiahousing offers you with an insight into Land Management.

Land information needs to be carefully managed to maximise its potential benefits. Over the last two decades, new capabilities for data collection and processing, together with expanding requirements of users, have directed awareness to the need for improved land information management strategies, concerned with the effective organisation of resources in order to achieve a set of objectives. These may include improvements to the coverage, content, compatibility, and reliability of information or access to it, and the possibility of integrating it with other data. The ultimate goal is to meet the needs of users more efficiently, effectively and equitably. Population growth, technological and social hazards, and environmental degradation have all to be taken into greater account today by policy makers, resource planners, and administrators who make decisions about the land. They need more detailed land information than has been traditionally available in order to render effective the professional supervision of landed property.

Land management refers to the way in which humans use the land, along with the plants and animals living on it, as a resource to fulfil the needs of society. The ways in which humans have utilised the land in Australia—the driest and least fertile inhabited continent of the world—have transformed markedly over the past 200 years.

Agriculture and Secondary Salinity

Agriculture can be defined as the process of cultivating plants and domesticating animals (livestock such as sheep, cattle and pigs) for the production of food and other goods intended for human consumption. Agriculture is one of the most vital national industries. It plays a key role in the nation's economic well being (through exports), as well as the domestic population's well being (through providing food).Land and water have been the basic elements of life support system on our planet since the dawn of civilisation. All great civilisations, flourished where these resources were available in plenty and they declined or perished with the depletion of these resources.

In recent years, the land resource has been subjected to a variety of pressures. Still it is surviving and sustaining mankind. What is alarming in the way land is being used is the tendency towards over-exploitation on account of a number of reasons leading this pristine resource being robbed of its resilience.

Of all the species on the earth, man is the chief culprit of this degradation. He views land in terms of its utility, meaning the capability to meet his perceived needs and wants. The most easily categorised varieties of land from the utility point of view are—land fit for use, land with potential for use and land which appears useless at least in the foreseeable future.

Here probably lies the genesis of the problem of land degradation and erosion of ecosystems. Mahatma Gandhi had said -"The Earth has enough for everybody's need but not for everybody's greed". Preserving, protecting and defending the land resources has been part of our age-old culture. The respect for the importance of land resources is best depicted in the conventional concept of Panchabhutas—land, water, fire, sky and air that constitute a set of divine forces. There are innumerable examples of the traditional conservation practices and systems, which are still surviving and are effective. But with the advent of modern age and the advent of newer forces, this tradition is fast deteriorating mainly on account of—consumerism, materialistic value systems, short-term profit-driven motives and greed of the users.

India constitutes 18 per cent of the world's population, 15 per cent of the live stock population and only 2 per cent of the geographic area, one per cent of the forest area and 0.5 per cent of pasture lands. The per capita availability of forests in India is only 0.08 per ha. as against the world average of 0.8 per cent , thus leading to the pressure on land and forests. This poses a major and urgent concern. In accordance with the National Remote Sensing Agency's (NRSA) findings there are 75.5 million ha. of wastelands in the country. In has been estimated that out of these around 58 million ha. are treatable and can be brought back to original productive levels through appropriate measures. At the moment, taking into account the efforts being made by all the various players in this field treating facilities are in place only for around 1 million ha. per year. At this rate, that there is no further degradation and also assuming that our efforts are 100 per cent successful, it will take around 58 years to complete the process.

Watershed degradation in the third world countries threatens the livelihood of millions of people and constraints the ability of countries to develop a healthy agricultural and natural resource base. Increasing population and livestock are rapidly depleting the existing natural resource base because the soil and vegetation system cannot support present level of use. As population continues to rise, the pressure on forests, community lands and marginal agricultural lands lead to inappropriate cultivation practices, forests removal and grazing intensities that leave a barren environment yielding unwanted sediment and damaging stream flow to down stream communities.

Watershed is a geo-hydrological unit which drains at a common point. Rains falling on the mountain start flowing down into small rivulets. Many of them, as they come down, join to form small streams. The small streams form bigger streams and then finally the bigger streams join to form a nallah to drain out of a village. The entire area that supplies water to a stream or river, i.e. the drainage basin or catchment area, is called the watershed of that particular stream or river.

Management of watershed thus entails the rational utilisation of land and water resources for optimum production but with minimum hazard to natural and human resources. The main objectives of watershed management are to protect the natural resources such as soil, water and vegetation from degradation. In the broader sense, it is an undertaking to maintain the equilibrium between elements of natural ecosystem of vegetation, land or water on the one hand and man's activities on the other hand.

When all possible inputs are obtaining, the man in the watershed still remains the most important component of the entire watershed system. The key issue is how far the people can be motivated, involved and organised to drive the movement. No significant improvement can be expected without the people being brought to centre-stage.

Role of Government

The Ministry of Rural Development has recently created a Department of Land Resources to act as a nodal department in the field of watershed management and development. This has the mandate of developing the valuable land resources of India, which are presently under various stages of degradation and it also endeavors to prevent further degradation of these resources through appropriate management and necessary measures.

The Department of Land Resources, being the nodal department has taken up certain new initiatives to play a more pro-active role in the Land Resource management in the country. At the conceptual level it has been realised that the management rather than the mere use of land is the central theme. There is no dearth of land, the real issue is management which should include: dynamic conservation, sustainable development and equitable access to the benefits of intervention.

The concept of sustainable development focuses on help for the very poor because they are left with no option but to destroy their own environment. It also includes the idea of cost-effective development using differing economic criteria to the traditional approach; that is to say development should not degrade environment quality, or reduce productivity in the long run. The greater issues of health control, appropriate technologies, food self-reliance, clean water and shelter for all are to be addressed. Sustainable development should seek to maintain an acceptable rate of growth in per capita real incomes without depleting the national capital asset stock or the natural environmental asset stock.

Equitable access to the benefits of development could be achieved either through land reforms or a dedicated and institutionalised mode of people's participation. Here, besides the Government, other players like the corporate sector, NGOs, various institutions and self-help groups can be involved.

Afforestation

Afforestation is the process of establishing a forest on land that is not a forest, or has not been a forest for a long time by planting trees or their seeds. The term may also be applied to the legal conversion of land into the status of royal forest.

The term reforestation generally refers to the reestablishment of the forest after its removal, for example from a timber harvest. Since the industrial revolution many countries have experienced centuries of deforestation, and some governments and non-governmental organisations directly engage in programs of afforestation to restore forests and assist in preservation of biodiversity.

The United States and northwestern Europe have more forest cover than at the beginning of the twentieth century. However, significant deforestation in South and Central America and in South Asia continues, although several nations such as Malaysia have worked hard to create a "green" environment.

In various arid, tropical, or sensitive areas, forests cannot re-establish themselves without assistance due to a variety of environmental factors. One of these factors is that, once forest cover is destroyed in arid zones, the land quickly dries out and becomes inhospitable to new tree growth. Other critical factors include overgrazing by livestock, especially animals such as goats, and over-harvesting of forest resources. Together these may lead to desertification and the loss of topsoil; without soil, forests cannot grow until the very long process of soil creation has been completed—if erosion allows this.

In some tropical areas, the removal of forest cover may result in a duricrust or duripan that effectively seal off the soil to water penetration and root growth. In many areas, reforestation is impossible above all because the land is in use by people. In these areas, reforestation requires the planting of tree seedlings, treeplanting. In other areas, mechanical breaking up of duripans or duricrusts is necessary, careful and continued watering may be essential, and special protection, such as fencing, may be required.

China has deforested the majority of its historical wooded areas. Although it has set official goals for re-forestation, these goals were set for a 80 year time horizon and are not significantly met by 2008. China has reached the point where timber yields have declined far below historic levels, due to overharvesting of trees beyond sustainable yield.

The European Union has paid farmers for afforestation since 1990, offering grants to turn farmland back into forest and payments for the management of forest. Between 1993 and 1997, EU afforestation policies made possible the re-forestation of over 5,000 square kilometres of land. A second program, running between 2000 and 2006, afforested in excess of 1000 square kilometres of land. A third such program began in 2007.

References

Conacher, Arthur; Conacher, Jeanette (1995). *Rural Land Degradation in Australia*. South Melbourne, Victoria: Oxford University Press Australia. p. 2.

Eswaran, H.; R. Lal and P.F. Reich. (2001). "Land degradation: an overview". *Responses to Land Degradation. Proc. 2nd. International Conference on Land Degradation and Desertification*. New Delhi, India: Oxford Press. Retrieved 2006-06-20.

Ian Sample (2007-08-31). "Global food crisis looms as climate change and population growth strip fertile land". *The Guardian*. Retrieved 2008-07-23.

Johnson, Douglas; Lewis, Lawrence. (2007), *Land Degradation; Creation and Destruction*, Maryland, USA

Stockings, Mike; Murnaghan, Niamh. (2000), *Land Degradation - Guidelines for Field Assesment*, Norwich, UK, pp. 7–15

6

Environmental Impact Assessment

An environmental impact assessment (EIA) is an assessment of the possible positive or negative impact that a proposed project may have on the environment, together consisting of the environmental, social and economic aspects.

The purpose of the assessment is to ensure that decision makers consider the ensuing environmental impacts when deciding whether to proceed with a project. The International Association for Impact Assessment (IAIA) defines an environmental impact assessment as "the process of identifying, predicting, evaluating and mitigating thebiophysical, social, and other relevant effects of development proposals prior to major decisions being taken and commitments made." EIAs are unique in that they do not require adherence to a predetermined environmental outcome, but rather they require decision -makers to account for environmental values in their decisions and to justify those decisions in light of detailed environmental studies and public comments on the potential environmental impacts of the proposal.

EIAs began to be used in the 1960s as part of a rational decision making process. It involved a technical evaluation that would lead to objective decision making. EIA was made legislation in the US in the National Environmental Policy Act (NEPA) 1969. It has since evolved as it has been used increasingly in many countries around the world. As per Jay et al.(2006), EIA as it is practiced today, is being used as a decision aiding tool rather than decision making tool. There is growing dissent on the use of EIA as its influence on development decisions is limited and there is a view it is falling short of its full potential.There is a need for stronger foundation of EIA practice through training for practitioners, guidance on EIA practice and continuing research.

EIAs have often been criticized for having too narrow spatial and temporal scope. At present no procedure has been specified for determining a system boundary for the assessment. The system boundary refers to 'the spatial and temporal boundary of the proposal's effects'. This boundary is determined by the applicant and the lead assessor, but in practice, almost all EIAs address the direct, on-site effects alone.

However, as well as direct effects, developments cause a multitude of indirect effects through consumption of goods and services, production of building materials and machinery, additional land use for activities of various manufacturing and industrial services, mining of resources etc. The indirect effects of developments are often an order of magnitude higher than the direct effects assessed by EIA. Large proposals such as airports or ship yards cause wide ranging national as well as international environmental effects, which should be taken into consideration during the decision-making process.

Broadening the scope of EIA can also benefit threatened species conservation. Instead of concentrating on the direct effects of a proposed project on its local environment some EIAs used a landscape approach which focused on much broader relationships between the entire population of a species in question. As a result, an alternative that would cause least amount of negative effects to the population of that species as a whole, rather than the local subpopulation, can be identified and recommended by EIA.

Environmental impact assessment, which is defined as an activity designed to identify and predict the impact on the biogeophysical environment and on man 's health and well-being of legislative proposals, policies, programmes, projects, and operational procedures, and to interpret and communicate information about the impacts. A number of terms have been used in English-speaking countries to distinguish (a) between natural and man-made environmental changes; and (b) between changes and the harmful and/or beneficial consequences of such changes. In one approach, a man-induced change is called an 'effect', while the harmful and/or beneficial consequences are called 'impacts'.

Sometimes, of course, an impact could be beneficial to some citizens but harmful to others. Another convention is to use the term 'impact' to denote only harmful effects. In still other countries, the words 'effects' and 'impacts' are synonymous and deleterious effects are termed 'damage'. No matter how the words are defined, however, a change/effect/impact is usually given in terms of its nature, its magnitude, and often its significance.

Methods to Carry out EIAs

There are various methods available to carry out EIAs, some are industry specific and some general methods:

— *Industrial products* - Product environmental life cycle analysis (LCA) is used for identifying and measuring the impact on the environment of industrial products. These EIAs consider technological activities used for various stages of the product: extraction of raw material for the product and for ancillary materials and equipment, through the production and use of the product, right up to the disposal of the product, the ancillary equipment and material.

— *Genetically modified plants* - There are specific methods available to perform EIAs of genetically modified plants. Some of the methods are GMP-RAM, INOVA etc.

— *Fuzzy Arithmetic* - EIA methods need specific parameters and variables to be measured to estimate values of impact indicators. However many of the environment impact properties cannot be measured on a scale e.g. landscape quality, lifestyle quality, social acceptance etc. and moreover these indicators are very subjective. Thus to assess the impacts we may need to take the help of information from similar EIAs, expert criteria, sensitivity of affected population etc. To treat this information, which is generally inaccurate, systematically, fuzzy arithmetic and approximate reasoning methods can be utilised. This is called as a fuzzy logic approach.

At the end of the project, an EIA should be followed by an audit. An EIA audit evaluates the performance of an EIA by comparing actual impacts to those that were predicted. The main objective of these audits is to make future EIAs more valid and effective. The two main considerations are:

— *scientific* - to check the accuracy of predictions and explain errors.

— *management*- to assess the success of mitigation in reducing impacts.

Some people believe that audits be performed as a rigorous scientific testing of the null hypotheses. While some believe in a simpler approach where you compare what actually occurred against the predictions in the EIA document.

After an EIA, the precautionary and polluter pays principles may be applied to prevent, limit, or require strict liability or insurance coverage to a project, based on its likely harms. Environmental impact assessments are sometimes controversial.

Environmental Impact Assessment Process

The EIA process makes sure that environmental issues are raised when a project or plan is first discussed and that all concerns are addressed as a project gains momentum through to implementation. Recommendations made by the EIA may necessitate the redesign of some project components, require further studies, suggest changes which alter the economic viability of the project or cause a delay in project implementation.

To be of most benefit it is essential that an environmental assessment is carried out to determine significant impacts early in the project cycle so that recommendations can be built into the design and cost-benefit analysis without causing major delays or increased design costs. To be effective once implementation has commenced, the EIA should lead to a mechanism whereby adequate monitoring is undertaken to realize environmental management. An important output from the EIA process should be the delineation of enabling mechanisms for such effective management.

The way in which an EIA is carried out is not rigid: it is a process comprising a series of steps. These steps are outlined below:

— screening

— scoping

— prediction and mitigation

— management and monitoring

— audit

In some cases, such as small-scale irrigation schemes, the transition from identification through to detailed design may be rapid and some steps in the EIA procedure may be omitted.

— Screening often results in a categorization of the project and from this a decision is made on whether or not a full EIA is to be carried out.

— Scoping is the process of determining which are the most critical issues to study and will involve community participation to some degree. It is at this early stage that EIA can most strongly influence the outline proposal.
— Detailed prediction and mitigation studies follow scoping and are carried out in parallel with feasibility studies.
— The main output report is called an Environmental Impact Statement, and contains a detailed plan for managing and monitoring environmental impacts both during and after implementation.
— Finally, an audit of the EIA process is carried out some time after implementation. The audit serves a useful feedback and learning function.

An EIA team for an irrigation and drainage study is likely to be composed of some or all of the following: a team leader; a hydrologist; an irrigation/drainage engineer; a fisheries biologist/ecologist; an agronomist/pesticide expert; a soil conservation expert; a biological/environmental scientist; an economist, a social scientist and a health scientist (preferably a epidemiologist). The final structure of the team will vary depending on the project. Specialists may also be required for fieldwork, laboratory testing, library research, data processing, surveys and modelling. The team leader will require significant management skill to co-ordinate the work of a team with diverse skills and knowledge.

There will be a large number of people involved in EIA apart from the full-time team members. These people will be based in a wide range of organizations, such as the project proposing and authorizing bodies, regulatory authorities and various interest groups. Such personnel would be located in various agencies and also in the private sector; a considerable number will need specific EIA training.

The length of the EIA will obviously depend on the programme, plan or project under review. However, the process usually lasts from between 6 and 18 months from preparation through to review. It will normally be approximately the same length as the feasibility study of which it should form an integral part. It is essential that the EIA team and the team carrying out the feasibility study work together and not in isolation from each other. This often provides the only opportunity for design changes to be made and mitigation measures to be incorporated in the project design.

The cost of the study will vary considerably and only very general estimates can be given here. Typically, costs vary from between 0.1 and 0.3 percent of the total project cost for large projects over US$ 100 million and from 0.2 to 0.5 percent for projects less than US$ 100 million. For small projects the cost could increase to between 1 and 3 percent of the project cost.

Screening

Screening is the process of deciding on whether an EIA is required. This may be determined by size (eg greater than a predetermined surface area of irrigated land that would be affected, more than a certain percentage or flow to be diverted or more than a certain capital expenditure). Alternatively it may be based on site-specific information. For example, the repair of a recently destroyed diversion structure is unlikely to require an EIA whilst a major new headwork structure may. Guidelines for whether or not an EIA is required will be country specific depending on the laws or norms in operation. Legislation often specifies the criteria for screening and full EIA. All major donors screen projects presented for financing to decide whether an EIA is required.

The output from the screening process is often a document called an Initial Environmental Examination or Evaluation (IEE). The main conclusion will be a classification of the project according to its likely environmental sensitivity. This will determine whether an EIA is needed and if so to what detail.

Scoping

Scoping occurs early in the project cycle at the same time as outline planning and pre-feasibility studies. Scoping is the process of identifying the key environmental issues and is perhaps the most important step in an EIA. Several groups, particularly decision makers, the local population and the scientific community, have an interest in helping to deliberate the issues which should be considered, and scoping is designed to canvass their views.

Scoping is important for two reasons. First, so that problems can be pinpointed early allowing mitigating design changes to be made before expensive detailed work is carried out. Second, to ensure that detailed prediction work is only carried out for important issues. It is not the purpose of an EIA to carry out exhaustive studies on all environmental impacts for all projects. If key issues are identified and a full scale EIA considered necessary then the scoping should include terms of reference for these further studies.

At this stage the option exists for cancelling or drastically revising the project should major environmental problems be identified. Equally it may be the end of the EIA process should the impacts be found to be insignificant. Once this stage has passed, the opportunity for major changes to the project is restricted.

Before the scoping exercise can be fully started, the remit of the study needs to be defined and agreed by the relevant parties. These will vary depending on the institutional structure. At a minimum, those who should contribute to determining the remit will include those who decide whether a policy or project is implemented, those carrying out the EIA (or responsible for having it carried out by others) and those carrying out parallel engineering and economic studies relating to the proposal. A critical issue to determine is the breadth of the study. For example, if a proposed project is to increase the area of irrigated agriculture in a region by 10%, is the remit of the EIA to study the proposal only or also to consider options that would have the same effect on production?

A major activity of scoping is to identify key interest groups, both governmental and non-governmental, and to establish good lines of communication. People who are affected by the project need to hear about it as soon as possible. Their knowledge and perspectives may have a major bearing on the focus of the EIA. Rapid rural appraisal techniques provide a means of assessing the needs and views of the affected population. The main EIA techniques used in scoping are baseline studies, checklists, matrices and network diagrams. These techniques collect and present knowledge and information in a straightforward way so that logical decisions can be made about which impacts are most significant.

Prediction and Mitigation

This stage forms the central part of an EIA. Several major options are likely to have been proposed either at the scoping stage or before and each option may require separate prediction studies. Realistic and affordable mitigating measures cannot be proposed without first estimating the scope of the impacts, which should be in monetary terms wherever possible. It then becomes important to quantify the impact of the suggested improvements by further prediction work. Clearly, options need to be discarded as soon as their unsuitability can be proved or alternatives shown to be superior in environmental or economic terms, or both. It is also important to test the "without project" scenario.

An important outcome of this stage will be recommendations for mitigating measures. This would be contained in the Environmental Impact Statement. Clearly the aim will be to introduce measures which minimize any identified adverse impacts and enhance positive impacts. Formal and informal communication links need to be established with teams carrying out feasibility studies so that their work can take proposals into account. Similarly, feasibility studies may indicate that some options are technically or economically unacceptable and thus environmental prediction work for these options will not be required.

Many mitigating measures do not define physical changes but require management or institutional changes or additional investment, such as for health services. Mitigating measures may also be procedural changes, for example, the introduction of, or increase in, irrigation service fees to promote efficiency and water conservation. By the time prediction and mitigation are undertaken, the project preparation will be advanced and a decision will most likely have been made to proceed with the project. Considerable expenditure may have already been made and budgets allocated for the implementation of the project.

Major changes could be disruptive to project processing and only accepted if prediction shows that impacts will be considerably worse than originally identified at the scoping stage. For example, an acceptable measure might be to alter the mode of operation of a reservoir to protect downstream fisheries, but a measure proposing an alternative to dam construction could be highly contentious at this stage. To avoid conflict it is important that the EIA process commences early in the project cycle.

This phase of an EIA will require good management of a wide range of technical specialists with particular emphasis on:

- prediction methods;
- interpretation of predictions, with and without mitigating measures;
- assessment of comparisons.

It is important to assess the required level of accuracy of predictions. Mathematical modelling is a valuable technique, but care must be taken to choose models that suit the available data. Because of the level of available knowledge and the complexity of the systems, physical systems are modelled more successfully than ecological systems which in turn are more successfully modelled than social systems. Social studies (including institutional capacity studies) will probably produce output in non-numerical terms. Expert advice, particularly from experts familiar with the locality, can provide quantification of impacts that cannot be modelled. Various techniques are available to remove the bias of individual opinion.

Checklists, matrices, networks diagrams, graphical comparisons and overlays, are all techniques developed to help carry out an EIA and present the results of an EIA in a format useful for comparing options. The main quantifiable methods of comparing options are by applying weightings, to environmental impacts or using economic cost-benefit analysis or a combination of the two. Numerical values, or weightings, can be applied to different environmental impacts to (subjectively) define their relative importance.

Assigning economic values to all environmental impacts is not recommended as the issues are obscured by the single, final answer. However, economic techniques, can provide insight into comparative importance where different environmental impacts are to be compared, such as either losing more wetlands or resettling a greater number of people.

When comparing a range of proposals or a variety of mitigation or enhancement activities, a number of characteristics of different impacts need to be highlighted. The relative importance of impacts needs agreeing, usually following a method of reaching a consensus but including economic considerations. The uncertainty in predicting the impact should be clearly noted. Finally, the time frame in which the impact will occur should be indicated, including whether or not the impact is irreversible.

Management and Monitoring

The part of the EIS covering monitoring and management is often referred to as the Environmental Action Plan or Environmental Management Plan. This section not only sets out the mitigation measures needed for environmental management, both in the short and long term, but also the institutional requirements for implementation. The term 'institutional' is used here in its broadest context to encompass relationships:

— established by law between individuals and government;
— between individuals and groups involved in economic transactions;
— developed to articulate legal, financial and administrative links among public agencies;
— motivated by socio-psychological stimuli among groups and individuals.

The above list highlights the breadth of options available for environmental management, namely: changes in law; changes in prices; changes in governmental institutions; and, changes in culture which may be influenced by education and information dissemination. All the management proposals need to be clearly defined and costed. One of the more straightforward and effective changes is to set-up a monitoring programme with clear definition as to which agencies are responsible for data collection, collation, interpretation and implementation of management measures.

The purpose of monitoring is to compare predicted and actual impacts, particularly if the impacts are either very important or the scale of the impact cannot be very accurately predicted. The results of monitoring can be used to manage the environment, particularly to highlight problems early so that action can be taken. The range of parameters requiring monitoring may be broad or narrow and will be dictated by the 'prediction and mitigation' stage of the EIA. Typical areas of concern where monitoring is weak are: water quality, both inflow and outflow; stress in sensitive ecosystems; soil fertility, particularly salinization problems; water related health hazards; equity of water distributions; groundwater levels.

The use of satellite imagery to monitor changes in land use and the 'health' of the land and sea is becoming more common and can prove a cost-effective tool, particularly in areas with poor access. Remotely sensed data have the advantage of not being constrained by political and administrative boundaries. They can be used as one particular overlay in a GIS. However, authorization is needed for their use, which may be linked to national security issues, and may thus be hampered by reluctant governments.

Monitoring should not be seen as an open-ended commitment to collect data. If the need for monitoring ceases, data collection should cease. Conversely, monitoring may reveal the need for more intensive study and the institutional infrastructure must be sufficiently flexible to adapt to changing demands. The information obtained from monitoring and management can be extremely useful for future EIAs, making them both more accurate and more efficient.

The Environmental Management Plan needs to not only include clear recommendations for action and the procedures for their implementation but must also define a programme and costs. It must be quite clear exactly how management and mitigation methods are phased with project implementation and when costs will be incurred. Mitigation and management measures will not be adopted unless they can be shown to be practicable and good value for money. The plan should also stipulate that if, during project implementation, major changes are introduced, or if the project is aborted, the EIA procedures will be re-started to evaluate the effect of such actions.

Auditing

In order to capitalise on the experience and knowledge gained, the last stage of an EIA is to carry out an Environmental Audit some time after completion of the project or implementation of a programme. It will therefore usually be done by a separate team of specialists to that working on the bulk of the EIA. The audit should include an analysis of the technical, procedural and decision-making aspects of the EIA. Technical aspects include: the adequacy of the baseline studies, the accuracy of predictions and the suitability of mitigation measures. Procedural aspects include: the efficiency of the procedure, the fairness of the public involvement measures and the degree of coordination of roles and responsibilities. Decision-making aspects include: the utility of the process for decision making and the implications for development, (adapted from Sadler in Wathern, 1988). The audit will determine whether recommendations and requirements made by the earlier EIA steps were incorporated successfully into project implementation. Lessons learnt and formally described in an audit can greatly assist in future EIAs and build up the expertise and efficiency of the concerned institutions.

Public Participation

Projects or programmes have significant impacts on the local population. Whilst the aim is to improve the well being of the population, a lack of understanding of the people and their society may result in development that has considerable negative consequences. More significantly, there may be divergence between national economic interests and those of the local population. For example, the need to increase local rice production to satisfy increasing consumption in the urban area may differ from the needs as perceived by the local farmers. To allow for this, public participation in the planning process is essential. The EIA provides an ideal forum for checking that the affected public have been adequately consulted and their views taken into account in project preparation.

The level of consultation will vary depending on the type of plan or project. New projects involving resettlement or displacement will require the most extensive public participation. As stated before, the purpose of an EIA is to improve projects and this, to some extent, can only be achieved by involving those people directly or indirectly affected. The value of environmental amenities is not absolute and consensus is one way of establishing values. Public consultation will reveal new information, improve understanding and enable better choices to be made. Without consultation, legitimate issues may not be heard, leading to conflict and unsustainability.

The community should not only be consulted they should be actively involved in environmental matters. The International Union for the Conservation of Nature, IUCN promotes the concept of Primary Environmental Care whereby farmers, for example, with assistance from extension services, are directly involved in environmental management. The earlier the public are involved, the better. Ideally this will be before a development proposal is fully defined. It is an essential feature of successful scoping, at which stage feedback will have the maximum influence. Openness about uncertainty should be a significant feature of this process. As the EIA progresses, public consultation is likely to be decreased though it is important to disseminate information. The publication of the draft Environmental Impact Statement (EIS), will normally be accompanied by some sort of public hearing that needs to be chaired by a person with good communication skills. He/she may not be a member of the EIA team.

There are no clear rules about how to involve the public and it is important that the process remains innovative and flexible. In practice, the views of people affected by the plan are likely to be heard through some form of representation rather than directly. It is therefore important to understand how decisions are made locally and what are the methods of communication, including available government extension services. The range of groups outside the formal structure with relevant information are likely to include: technical and scientific societies; Water User Groups; NGOs; experts on local culture; and religious groups. However, it is important to find out which groups are under-represented and which ones are responsible for access to natural resources, namely: grazing, water, fishing and forest products. The views of racial minorities, women, religious minorities, political minorities and lower cast groups are commonly overlooked.

There has been an enormous increase in the number of environmental NGOs and "Green" pressure groups throughout the world. Such organizations often bring environmental issues to the attention of the local press. However, this should not deter consultation with such organizations as the approach to EIA should be open and positive with the aim of making improvements. Relevant NGOs should be identified and their experience and technical capacity put to good use.

In some countries, open public meetings are the most common technique to enable public participation. However, the sort of open debate engendered at such meetings is often both culturally alien and unacceptable. Alternative techniques must be used. Surveys, workshops, small group meetings and interviews with key groups and individuals are all techniques that may be useful. Tools such as maps, models and posters can help to illustrate points and improve communication. Where resettlement is proposed, extensive public participation must be allowed which will, at a minimum, involve an experienced anthropologist or sociologist who speaks the local language. He/she can expect to spend months, rather than weeks, in the field.

Information dissemination can be achieved using a number of mechanisms including the broadcasting media, in particular newspapers and radio. Posters and leaflets are also useful and need to be distributed widely to such locations as schools, clinics, post offices, community centres, religious buildings, bus stops, shops etc. The EIA process must be seen to be fair. The public participation/consultation and information dissemination activities need to be planned and budgeted. The social scientist team member should define how and when activities take place and also the strategy: extensive field work is expensive.

Uncertainty

An EIA involves prediction and thus uncertainty is an integral part. There are two types of uncertainty associated with environmental impact assessments: that associated with the process and, that associated with predictions. With the former the uncertainty is whether the most important impacts have been identified or whether recommendations will be acted upon or ignored. For the latter the uncertainty is in the accuracy of the findings. The main types of uncertainty and the ways in which they can be minimized are discussed by de Jongh in Wathern. They can be summarized as follows:

— uncertainty of prediction: this is important at the data collection stage and the final certainty will only be resolved once implementation commences. Research can reduce the uncertainty;

— uncertainty of values: this reflects the approach taken in the EIA process. Final certainty will be determined at the time decisions are made. Improved communications and extensive negotiations should reduce this uncertainty;

— uncertainty of related decision: this affects the decision making element of the EIA process and final certainty will be determined by post evaluation. Improved coordination will reduce uncertainty.

The importance of very wide consultation cannot be overemphasized in minimizing the risk of missing important impacts. The significance of impacts is subjective, but the value judgements required are best arrived at by consensus: public participation and consultation with a wide sector of the community will reduce uncertainty. One commonly recurring theme is the dilemma of whether to place greater value on short-term benefits or long-term problems.

The accuracy of predictions is dependent on a variety of factors such as lack of data or lack of knowledge. It is important not to focus on predictions that are relatively easy to calculate at the expense of impacts that may be far more significant but difficult to analyse. Prediction capabilities are generally good in the physical and chemical sciences, moderate in ecological sciences and poor in social sciences. Surveys are the most wide-spread technique for estimating people's responses and possible future actions.

The results of the EIA should indicate the level of uncertainty with the use of confidence limits and probability analyses wherever possible. Sensitivity analysis similar to that used in economic evaluation, could be used if adequate quantifiable data are available. A range of outcomes can be found by repeating predictions and adjusting key variables.

EIA cannot give a precise picture of the future, much as the Economic Internal Rate of Return cannot give a precise indication of economic success. EIA enables uncertainty to be managed and, as such, is an aid to better decision making. A useful management axiom is to preserve flexibility in the face of uncertainty.

EIA Techniques

- Baseline studies
- The ICID Check-list
- Matrices
- Network diagrams
- Overlays
- Mathematical modelling
- Expert advice
- Economic techniques

Baseline Studies

Baseline studies using available data and local knowledge will be required for scoping. Once key issues have been identified, the need for further in-depth studies can be clearly identified and any additional data collection initiated. The ICID Check-list will be found useful to define both coarse information required for scoping and further baseline studies required for prediction and monitoring. Specialists, preferably with local knowledge, will be needed in each key area identified. They will need to define further data collection, to ensure that it is efficient and targeted to answer specific questions, and to quantify impacts. A full year of baseline data is desirable to capture seasonal effects of many environmental phenomena. However, to avoid delay in decision making, short-term data monitoring should be undertaken in parallel with long-term collection to provide conservative estimates of environmental impacts.

ICID Check-list

A comprehensive and user-friendly checklist is an invaluable aid for several activities of an EIA, particularly scoping and defining baseline studies. "The ICID Environmental Check-List to Identify Environmental Effects of Irrigation, Drainage and Flood Control Projects" is recommended for use in any irrigation and drainage EIA. The Check-list has been prepared for non-specialists and enables much time-consuming work to be carried out in advance of expert input. It includes extensive data collection sheets.

The very simple layout of the sheet enables an overview of impacts to be presented clearly which is of enormous value for the scoping process. Similarly, data shortages can be readily seen. The process of using the ICID Check-list may be repeated at different stages of an EIA with varying levels of detail. Once scoping has been completed, the results sheet may be modified to omit minor topics and to change the horizontal classification to provide further information about the impacts being assessed. At this point the output from the Check-list can be useful as an input to matrices. The ICID Check-list is also available as a WINDOWS based software package. This enables the rapid production of a report directly from the field study.

Matrices

The major use of matrices is to indicate cause and effect by listing activities along the horizontal axis and environmental parameters along the vertical axis. In this way the impacts of both individual components of projects as well as major alternatives can be compared. The simplest matrices use a single mark to show whether an impact is predicted or not. However it is easy to increase the information level by changing the size of the mark to indicate scale, or by using a variety of symbols to indicate different attributes of the impact. The choice of symbols in this example enables the reader to see at a glance whether or not there was an impact and, if so, whether the impact was beneficial or detrimental, temporary or permanent.

ICOLD has prepared a large and comprehensive matrix for use in EIAs for dams. The system of symbols for each box shows: whether the impact is beneficial or detrimental; the scale of the impact; the probability of occurrence; the time-scale of occurrence; and, whether the design has taken the impact into account. This comprehensive approach, however, makes the final output rather difficult to use and a maximum of three criteria is recommended per impact to maintain clarity. The most important criteria are: magnitude, or degree of change; geographical extent; significance; and, special sensitivity. "Significance" could be further sub-divided to indicate why an impact is significant. For example, it may be because of irreversibility, economic vulnerability, a threat to rare species etc. "Special sensitivity" refers to locally important issues. A series of matrices at all stages of the EIA process can be a particularly effective way of presenting information. Each matrix may be used to compare options rated against a few criteria at a time.

The greatest drawback of matrices are that they can only effectively illustrate primary impacts. Network diagrams, described below, are a useful and complementary form of illustration to matrices as their main purpose is to illustrate higher order impacts and to indicate how impacts are inter-related. Matrices help to choose between alternatives by consensus. One method is to make pair-wise comparisons. It provides a simple way for a group of people to compare a large number of options and reduce them to a few choices. First a matrix is drawn with all options listed both horizontally and vertically. Each option is then compared with every other one and a score of 1 assigned to the preferred option or 0.5 to both options if no preference is agreed.

Network diagrams

A network diagram is a technique for illustrating how impacts are related and what the consequences of impacts are. For example, it may be possible to fairly accurately predict the impact of increased diversions or higher irrigation efficiencies on the low flow regime of a river. However, there may be many and far reaching secondary or tertiary consequences of a change in low flow. These consequences can be illustrated using network diagrams. For example, reduced low flows are likely to reduce the production of fish which may or may not be of importance depending on the value (either ecological or economic) of the fish. If fish are an important component of diet or income, the reduction may lead to a local reduction in the health status, impoverishment and possibly migration. Also, reduced low flow coupled with increased pollution, perhaps as a result of increased agricultural industry, may further damage the fish population as well as reduce access to safe water.

Overlays

Overlays provide a technique for illustrating the geographical extent of different environmental impacts. Each overlay is a map of a single impact. For example, saline effected areas, deforested areas, limit of a groundwater pollution plume etc can be analysed and clearly demonstrated to non experts. The original technique used transparencies which is somewhat cumbersome.

However, the development of Geographic Information Systems (GIS) can make this technique particularly suitable for comparing options, pinpointing sensitive zones and proposing different areas or methods of land management.

Mathematical Modelling

Mathematical modelling is one of the most useful tools for prediction work. It is the natural tool to assess both flow quantities and qualities (eg salt/ water balances, pollution transport, changing flood patterns). However, it is essential to use methods with an accuracy which reflects the quality of the input data, which may be quite coarse. It should also be appreciated that model output is not necessarily an end in itself but may be an input for assessing the impact of changes in economic, social and ecological terms. Mathematical modelling was used very effectively to study the Hadejia-Jama' are region in Nigeria. In this case the modelling demonstrated the most effective method of operating upstream reservoirs in order to conserve economically and socially valuable, and ecologically important downstream wetlands. Optimal operation was found to be considerably different from the traditional method originally proposed. Under the revised regime the economic returns were also found to be higher.

Expert Advice

Expert advice should be sought for predictions which are inherently non-numeric and is particularly suitable for estimating social and cultural impacts. It should preferably take the form of a consensus of expert opinion. Local experience will provide invaluable insight. Expert opinions are also likely to be needed to assess the implications of any modelling predictions. For example, a model could be developed to calculate the area of wetlands no longer annually flooded due to upstream abstractions. However, the impact on wetland species or the reduction in wetland productivity resulting from the reduced flooding may not be so precisely quantifiable but require a prediction based on expert opinion.

Economic Techniques

Economic techniques have been developed to try to value the environment and research work is continuing in environmental economics. The most commonly used methods of project appraisal are cost-benefit and cost-effectiveness analysis. It has not been found easy to incorporate environmental impacts into traditional cost-benefit analysis, principally because of the difficulty in quantifying and valuing environmental effects. An EIA can provide information on the expected effects and quantify, to some extent, their importance. This information can be used by economists in the preparation of cost-benefit calculations. Cost effectiveness analysis can also be used to determine what is the most efficient, least-cost method of meeting a given environmental objective; with costs including forgone environmental benefits. However, defining the objective may not be straightforward.

Valuing the environment raises complex and controversial issues. The environment is of value to the actual users (such as fishermen), to potential users (future generations or migrants), and to those who do not use it but consider its existence to have an intrinsic value (perhaps to their "quality of life"). Clearly it is difficult to quantify such values. Nevertheless, attempts have been made and the two most useful methods for irrigation projects in developing countries are "Effect on Production" (EOP) and "Preventive Expenditure and Replacement Costs" (PE/RC).

The EOP method attempts to represent the value of change in output that results from the environmental impact of the development. This method is relatively easy to carry out and easily understood. An example would be the assessment of the reduced value of fish catches due to water pollution or hydrological changes. The PE/RC method makes an assessment of the value that people place on preserving their environment by estimating what they are prepared to pay to prevent its degradation (preventive expenditure) or to restore its original state after it has been damaged (replacement cost). Both methods have weaknesses and must be used judiciously.

Environmental health effects present similar problems, cost-effectiveness analysis is a useful tool in the selection of mitigating or control measures, but for ex-ante project appraisal the incompatibility of human health and monetary values has forced economists to develop other techniques and indicators. A recent publication by Phillips et al. deals with the principles and methods of cost-effectiveness analysis and its application to decisions about the control of vector-borne diseases, particularly the control of disease vectors.

EIA is often referred to as an Environmental Impact Statement (EIS). In addition to summarizing the impacts of the alternatives under study this report must include a section on follow up action required to enable implementation of proposals and to monitor long-term impacts. The purpose of an EIA is not to reach a decision but to present the consequences of different choices of actions and to make recommendations to a decision maker.

EIA and Sustainable Development

Environmental Impact Assessment (EIA) has a history of over two decades. The philosophy and practice of EIA have been incorporated into legislative and administrative systems across the world. Its acceptance does not only reflect the desire and necessity to integrate environmental considerations into decision-making, but also the relative simplicity and flexibility of the approach. However, this integration has been limited, both in terms of the scope of application and in the effectiveness of practice. The international activity related to the promotion of environmental protection has initiated new management tools, such as environmental auditing, product assessment and life-cycle analysis. Existing tools, primarily developed for other purposes, such as risk assessment and cost-benefit analysis are widely used for environmental decision-making.

Basically, EIA is an iterative assessment and decision process, rather than a specific technique, which attempts to determine the impacts of policies and/or activities on the environment so that there is an opportunity for interested parties to decide whether those impacts are acceptable. The definition of the environment includes the receiving environmental media and their physical components, living species occupying these media, and the built, cultural, and social environment. EIA is not a formal decision-making process, rather it is a management tool. It has several objectives which relate to the identification of potential problems in the decision process, the provision for the balancing of costs and benefits, the reduction of unacceptable impacts, and the provision of interdisciplinary inputs to environmental decisions.

Management decisions require the status quo to be a relevant consideration, and the "to action" alternative is an important consideration in the EIA process. There is no generally accepted definition of the purpose and nature of EIA. It is a term which has developed over the years, in the light of environmental concerns, policy, assessment techniques, and practice. The lack of a common definition can be seen as a disadvantage, offering different approaches between different countries and agencies, and a lack of clarity as to the requirements for potential users. Certainly EIA must be a dynamic process, constantly under review to ensure that its purpose and application is directly relevant to the needs of the time.

A sustainable process or condition is one that can be maintained indefinitely without progressive diminution of valued qualities inside or outside the system in which the process operates or the condition prevails. The proposition that particular human practices would prove unsustainable has cropped up in literature going all the way back to the ancient Greeks and somewhat more frequently and sweepingly in the two hundred years since the work of Malthus, above all in the period since World War II. Only in the past five years, however, has sustainability become a catchword capable of capturing the attention not only of environmental scientists and activists but also of (some) mainstream economists, other social scientists, and policymakers.

This enhanced salience presumably resulted from a suite of coincident factors. For one, the world community is no longer transfixed by the Cold War. A second factor is the reluctant appreciation of the severity of the debt crisis in the developing world. A third is the substantial advancement in scientific understanding of the magnitude and consequences of ongoing global environmental transformations, including the depletion of stratospheric ozone, the buildup of greenhouse gases, and the destruction of biodiversity. Also very important has been the attention given to the notion of sustainable development in the report of the World Commission on Environment and Development and the avalanche of related studies that has followed.

Notwithstanding the extraordinary growth of the "sustainability" literature in the past few years (an unsustainable process, to be sure!), much of the analysis and discussion of this topic remains mired in terminological and conceptual ambiguities, as well as in disagreements about facts and practical implications. These problems arise in part because the sustainability of the human enterprise in the broadest sense depends on technological, economic, political, and cultural factors as well as on environmental ones and in part because practitioners in the different relevant fields see different parts of the picture, typically think in terms of different time scales, and often use the same words to mean different things.

It is therefore appropriate, even though this introductory chapter and the conference of which it was originally a part are supposed to focus on the biogeophysical aspects of sustainability, to begin by locating the biogeophysical aspects within the context of the wider debate about what sustainability means and implies. Some problems with defining biogeophysical sustainability in practical terms, the connection between biogeophysical sustainability and related concepts such as carrying capacity and the distinction between renewable and nonrenewable resources, the state of knowledge and debate about the character and origins of threats to biogeophysical sustainability, and some implications of the current state of knowledge and ignorance of these matters.

Biogeophysical Sustainability

Much of the current salience of concepts of sustainability has come from a wide-ranging international discussion about sustainable development, which has been defined variously as, for example:

— Meeting the needs of the present without compromising the ability of future generations to meet their own needs

— Improving the quality of human life while living within the carrying capacity of supporting ecosystems

— Economic growth that provides fairness and opportunity for all the world's people, not just the privileged few, without further destroying the world's finite natural resources and carrying capacity.

If improvements in the human condition are to be not only achieved but also sustained, all of the ills will need to be addressed; this is so because failure to address any one of them can eventually undermine the progress made on all the others. As the human enterprise expands, interdependencies mediated through the world economy, the global environmental commons, and international political and military relations link and intensify the threats posed by each of these ills. Thus the requires meets for sustainability include not only the environmental factors to which we will shortly turn in detail but also military, political, and economic ones.

The environmental aspect of sustainability has been the subject of a rich literature, albeit only recently with the term sustainability appearing explicitly. As with the concept of sustainable development, however, the definitions of environmental sustainability to be found in the literature recent enough to use that term are often circular or unsatisfying in other ways. Consider the following capsule definitions:

— Sustainability refers to a process or state that can be maintained indefinitely

— Natural resources must be used in ways that do not create ecological debts by overexploiting the carrying and productive capacity of the Earth

— A minimum necessary condition for sustainability is the maintenance of the total natural capital stock at or above the current level.

The first statement is essentially a dictionary definition of sustainability; it tells us only what we already knew sustainability to mean. The second statement introduces the interesting term "ecological debt" to describe an element of unsustainability, but the elaboration in terms of overexploiting carrying capacity and productive capacity is not much help, insofar as it merely transfers the definitional burden to over-exploitation and carrying capacity. The third statement offers an actual specification of at least one element of sustainability, but there is still buried within it a definitional problem: How is "total natural stock" to be defined and measured? Assuming this hurdle can be overcome, the further question will surely arise: What is inviolable about the current level? Can environmental scientists give a good answer?

Of course, all serious writers on environmental sustainability go beyond the sorts of capsule definitions cited above and elaborate what sustainability might entail and require. The 1980 World Conservation Strategy of the International Union for the Conservation of Nature, the United Nations Environment Program, and the World Wildlife Fund concludes, for example, that sustainability requires "maintenance of essential ecological processes and life-support systems; preservation of genetic diversity; and sustainable utilization of species and resources." This three-part prescription seems to consist of different facets of the same thing: preservation of genetic diversity and sustainable use are essential to maintain essential ecological processes and life support systems.

Environmental issues have become so popular that politicians around the world no longer need to be persuaded of their importance. Natural scientists have been using the word sustainability for a fairly long time, and recently social scientists as well as politicians have started to use it quite frequently. However, it has yet to be defined clearly. Recommendations have been made for the definition of measurements and indicators of sustainability. Although by no means final, the following working definition of biophysical sustainability is satisfactory for the time being: Biophysical sustainability means maintaining or improving the integrity of the life support system of Earth. Sustaining the biosphere with adequate provisions for maximizing future options includes enabling current and future generations to achieve economic and social improvement within a framework of cultural diversity while maintaining (a) biological diversity and (b) the biogeochemical integrity of the biosphere by means of conservation and proper use of air, water, and land resources. Achieving these goals requires planning and action at local, regional, and global levels and specifying short- and long-term objectives that allow for the transition to sustainability.

Biophysical refers not only to biology and physics but also to geology and chemistry. This is expressed in the definition, particularly through mention of biogeochemical integrity. Natural science has become so interdisciplinary that it is often confusing; nevertheless physics, chemistry, geology, and biology remain the most basic disciplines. Biogeophysicochemistry expresses them all in one word, albeit an exceptionally long one. Defining terms such as sustainability and sustainable development with reference to the global environment is, to my mind, complicated by the fact that humanity has been considered special and separate from other animals and plants. This has not always been the case.

The Earth is divided into three spheres: atmosphere, hydrosphere, and lithosphere. The biosphere was added later as the fourth sphere but, unlike the others, includes those parts of the atmosphere, hydrosphere, and lithosphere in which life exists. Plants and animals are, of course, part of the biosphere, but more importantly, humans are included as just one species of animal and are not treated specially. In recent years, particularly when serious environmental problems were recognized, human activity was so intense and pervasive that it came to be considered—for example, by the Man and the Biosphere Programme as separate from the activity of other forms of life.

Biophysical sustainability must, therefore, mean the sustainability of the biosphere minus humanity. Humanity's role has to be considered separately as economic or social sustainability. Likewise, sustainable development should mean both sustainability of the biophysical medium or environment and sustainability of human development, with the latter sustaining the former.

Social, political, and cultural issues come into play in a number of ways at two critical stages in any discussion of environmental sustainability.

Stage 1. In Deciding

— What is to be sustained? That is, what relative ranking is to be given to, say, current resource productivity, productive potential, or genetic diversity?

— What attributes, or combinations of attributes, of a particular system are to be maintained nondecreasing: average productivity, stability, resilience, or adaptability?

— Over what time scale is this sustenance desired?

— Who is to benefit? If a tradeoff is necessary between current and future consumption and well-being, or between the well-being of one community and that of another, who is to decide and how?

— Should it be economic value of any resource flow or stock that is maintained non-decreasing, or should it be the physical quantity of that flow?

Stage 2. In understanding

— Why is there environmental unsustainability, however defined, in the world today?

— How would one achieve or move toward whatever notion of an environmentally sustainable society that is decided on in stage 1?

Stage 1 requires an explication of differing individual and cultural values, preferences, as well as beliefs about and approaches to a highly uncertain and unknowable future and then the resolution of such differences through some social process. Stage 2 requires an understanding of the complex array of social, political, and cultural factors in today's world that lead to environmentally unsustainable behavior.

Once this is clearly realized, it is easier to understand where our contributions as biophysicists and ecologists can be and ought to be in informing the process of reaching some societal consensus on the issues in stage 1. We have often made implicit decisions about the issues raised in Stage 1. We should therefore proceed as follows:

1. Clearly state the assumptions are making about reality in a particular case, examine whether some assumptions are commonly shared, and determine the extent to which these may be justified. For instance, perhaps most ecologists believe that whatever is to be maintained nondecreasing in an ecosystem should be measured in physical, not economic, terms. This follows from their rejection of the belief commonly held and vigorously promoted by most economists: that technological change can continuously compensate for reduction in physical resource flows, thus preventing utility from de c
2. Clearly state what value-based choices of objectives, of their ranking, of time horizons, and of users are being implicitly made in any particular case.
3. Identify a few scenarios corresponding to choices different from those that we might want to make.
4. Synthesize the current state of knowledge about the relationships between biophysical processes that affect different objectives at different temporal and spatial scales. That is, what intensity of harvesting under what technique of logging can be maintained in a tropical forest at a nondecreasing level for what time period? What would the implications of a nondecreasing resilience requirement be?
5. Identify a sparse set of indicators that best relate to each combination of objective, scale, and so forth and possibly identify threshold values for them. For instance, what would be the best indicator of stable harvests in the above-mentioned forest? What would be the indicator of resilience in the same system? What scales (spatial and temporal) may be most appropriate or sensible for measuring what attribute or type of sustainability?
6. Explore the ways in which the different scenarios interact; that is, the synergisms and contradictions among objectives, attributes, and indicators and between sustainability in general and other societal objectives. What are the tradeoffs between, say, maintaining timber productivity and maintaining biodiversity in a forest, or between average production and resilience? What are the tradeoffs between different levels of these attributes of sustainability and between the net yield or human wellbeing produced and the manner in which it is distributed within society?

The 1991 "Strategy for Sustainable Living" by the same triad of organizations says that "sustainable use means use of an organism, ecosystem, or other renewable resource at a rate within its capacity for renewal." Operating within the capacity for renewal clearly is one of the key elements of sustainability, but this formulation does not deal with either nonrenewable resources or the possible off-site, out-of-ecosystem impacts through which exploitation of one resource within its capacity for renewal might adversely affect the renewability of other resources or the sustainability of other ecosystems.

The economist Herman Daly, who has been a pioneer in thinking systematically about these matters recently offered a more helpful three-part specification of the ingredients of sustainability:

— Rates of use of renewable resources do not exceed regeneration rates
— Rates of use of nonrenewable resources do not exceed rates of development of renewable substitutes
— Rates of pollution emission do not exceed assimilative capacities of the environment.

The first of these conditions, by being stated in the aggregate, partly addresses the problem of off-site impacts associated with the exploitation of individual renewable resources: the regeneration rates constraints presumably reflect cross-resource or cross-ecosystem impacts occurring within the overall pattern of resource exploitation.

The second condition, the rate of use of nonrenewable resources, offers a clever solution to the question of how any use of nonrenewable resources can be contemplated within a sustainability framework. Daly offers a detailed formulation on how to ensure that this condition is met, by earmarking part of the proceeds from the exploitation of nonrenewable resources for the development of renewable alternatives.

Daly's third condition, on rates of pollution emission, does not seem as satisfying. If assimilative capacity of the environment means the capacity to assimilate the pollution without any adverse effect on human health or welfare (including through diminution of ecosystem services), the difficulty is that there are many kinds of pollution for which the assimilative capacity, so defined, is probably zero (including, for example, ionizing radiation, chlorofluorocarbons, lead, and more). It does not seem to insist on no harm from pollution as a condition of sustainability; the question is rather what level of harm is tolerable on a steady-state basis, in exchange for the benefits of the activity that produces the harm.

We would also add to Daly's three-part formulation that the first condition applies to resources for which substitution at the required scale is currently and foreseeably impossible (essential resources). It is useful to distinguish those from resources for which substitutes are currently or foreseeably available (substitutable resources). Renewable substitutable resources could be sustainably exhausted on the same basis as nonrenewable substitutable resources.

Biogeophysical Sustainability

The two most important questions relating to a definition of biogeophysical sustainability are "What is to be sustained?" and "For how long?" It is useful to distinguish, with respect to these questions, between what one would like the answers to be in theory and what one might have to settle for in practice.

Several practical problems intrude on the attractiveness of this theoretical approach. First, even the existing benefit flows from the environment—not to mention the potentially continuously derivable benefit flows are partly unknown (indeed, partly unknowable) and also partly incommensurable.

Second, insisting that potential benefit flows remain constant over very long periods of time is problematic because environmental conditions and processes—climate, topography, the biota—are occurring all the time even in the absence of human interventions. The potential magnitudes of such changes over the very long term make the concept of forever essentially meaningless, at least in relation to the sustainability of conditions that humans of today care about.

Third, it is conceivable that technological improvements will permit well-being to be maintained despite diminished benefit flows from the environment. This argument is probably the one most heavily relied upon by those not convinced of the need to maintain the stream of environmental services undiminished. But attempts to substitute technology for diminishing or otherwise inadequate environmental services invariably entail monetary costs and often generate significant new environmental impacts. In some cases, these additional costs and impacts may more than offset the (presumed) benefits of the activities that necessitated augmentation of the natural environmental services in the first place; and even if it is supposed that this will not be the case, it strikes us as imprudent in the extreme to assume that suitable technology for replacing whatever environmental services are lost will become available in a timely manner and on the requisite scale.

In any case, in light of the difficulties of measuring actual and potential environmental benefit flows, and in light of the conceptual and practical problems of insisting on no degradation forever, it may be necessary in practice to settle for trying to sustain the magnitude and quality of environmental stocks. The time scale on which this ought to be ensured might be defined in practical terms by a resource or stock half-life of 500 to 1,000 years, a period much longer than current planning horizons, but much shorter than geologic time. A tentative rule for prudent practice, then, would be to constrain the degradation of monitorable environmental stocks to not more than 10 percent per century.

This prudent-practice approach, because that is what can most easily be measured. This sidesteps slightly the argument with the economists and technologists over what is so special about the current levels; putting the argument in terms of degradation rates relies on the presumed circumstance that there is some degradation rate that is too high to be regarded as sustainable, even allowing for economic substitution and technological change.

Of course, it would not really be acceptable to run down environmental stocks at 10 percent a century indefinitely. The point is rather that a rate of 10 percent a century (which after all means about 0.1 percent a year) is slow enough to give society a reasonable chance of figuring out what this degradation is costing, which forms of degradation can be compensated for, how those forms can be stopped that cannot be compensated for or tolerated, and so on, before it is too late. At current degradation rates, by contrast, which are typically an order of magnitude or so higher (that is, in the range of 100 percent a century or more), natural services will be devastated before society even understands what is happening, let alone finds time to take evasive action on the needed scale.

A more important source of disagreement, however, are the differing assumptions, perceptions, and knowledge about (a) the importance of environmental conditions and processes in supporting human well-being, (b) the sensitivity of those conditions and processes to disruption, and (c) the character and amenability of society to remedy the anthropogenic impacts now threatening such disruption. Confusion about the sensitivity of those conditions and processes to disruption is evident in the comment attributed to economist William Nordhaus that only 3 percent of gross national product (GNP) in the United States depends on the environment. In fact, the entire GNP in the US. depends, ultimately, on maintaining the biophysical requisites of sustainability. Furthermore, the importance of agriculture (the economic sector to which Nordhaus apparently was referring) is vastly underestimated by its present share of GNP.

The greatest disparities in interpretation of the relationships between the human enterprise and Earth's life support systems seem, in fact, to be those between ecologists and economists. Members of both groups tend to be highly self-selected and to differ in fundamental worldviews. Most ecologists have a passion for the natural world, where the existence of limits to growth and the consequences of exceeding those limits are apparent. Ecologists recognize that a unique combination of highly developed manual dexterity, language, and intelligence has allowed humanity to increase vastly the capacity of the planet to support Homo sapiens; nonetheless, they perceive humans as being ultimately subject to the same sorts of biophysical constraints that apply to other organisms.

Economists, in contrast, tend to receive little or no training in the physical and natural sciences. Few explore the natural world on their own, and few appreciate the extreme sensitivity of organisms—including those upon which humanity depends for food, materials, pharmaceuticals, and free ecosystem services—to seemingly small changes in environmental conditions. Most treat economic systems as though they were completely disconnected from the planet's basic life support systems. The narrow education and inclinations of economists in these respects are thus a major source of disagreements about sustainability.

Some of the responsibility for these continuing disagreements also rests, however, on the failure of ecologists and other environmental scientists to make a case for the importance of environmental conditions and processes and for the magnitude of anthropogenic threats to these, in terms understandable by and persuasive to others. This problem is partly a matter of too few environmental scientists having made the effort to articulate a coherent case, but also partly a matter of the great gaps in the environmental science itself. Nor has it helped that environmental scientists are often as ignorant about economic principles, and their relevance to environmental protection, as economists are about ecological principles.

Approaching consensus about biogeophysical sustainability clearly will require more research, more communication across disciplines, and more education of the public and policymakers about a multitude of issues, notably:

— The character and dimensions of the ways environmental structure and function affect human well-being, including the identification of environmental services and the quantification of their magnitude and value.

— The ways human impacts imperil environmental services, involving identification of environmental systems at risk and the causes, extent, time scales, and degree of irreversibility of anthropogenic threats to these systems.

— The amenability of the threats to remedy, including potential improvements in technology and management, use of economic incentives to induce appropriate changes, and the social, political, and economic barriers to implementation of the remedies.

References

Carroll, B. and Turpin T. (2009). *Environmental impact assessment handbook, second edition* Thomas Telford Ltd.

Glasson, J; Therivel, R; Chadwick A, (2005). *Introduction to Environmental Impact Assessment*, Routledge, London.

Hanna, K; (2009). *Environmental Impact Assessment: Practice and Participation* Second edition, Oxford,

Petts, J. (ed), *Handbook of Environmental Impact Assessment* Vol 1 & 2, Blackwell, Oxford.

7

Conservation and Protection of Environment

By now, all of us have realized how important it is to protect the environment for our own survival. The term 'conservation' of environment relates to activities which can provide individual or commercial benefits, but at the same time, prevent excessive use leading to environmental damage. Conservation may be distinguished from preservation, which is considered to be "maintaining of nature as it is, or might have been before the intervention of either human beings or natural forces." We know that natural resources are getting depleted and environmental problems are increasing. It is, therefore, necessary to conserve and protect our environment.

Conservation Ecology

The term conservation ecology covers a very wide range of subjects, basically, any part of ecology that has a bearing on conservation. The terms Ecology and Conservation are frequently used interchangeably, but, although ecologists are frequently also conservationists, this does not necessarily have to be the case.

The first principle of ecology is that each living organism has an ongoing and continual relationship with every other element that makes up its environment. An ecosystem can be defined as any situation where there is interaction between organisms and their environment. The ecosystem is composed of two entities, the entirety of life, the biocoenosis and the medium that life exists in the biotope. Within the ecosystem, species are connected and dependent upon one another in the food chain, and exchange energy and matter between themselves and with their environment.

Conservation biology, or conservation ecology, is the science of analysing and protecting Earth's biological diversity. Conservation biology is an interdisciplinary subject drawing on biological, physical and social sciences, economics, and the practice of natural-resource management. The rapid decline of biological systems around the world means that conservation biology is often referred to as a "Discipline with a deadline". Conservation ecology addresses population dynamics issues associated with the small population sizes of rare species (e.g., minimum viable populations).

The term "conservation biology" refers to the Conservation biology is the scientific study of the phenomena that affect the maintenance, loss, and restoration of biological diversity and the application of science to the conservation of genes, populations, species, and ecosystems. For the history of biodiversity conservation and volunteer activity, see conservation movement.

In the 19th century actions in the United Kingdom, the United States and certain other western countries emphasized the protection of habitat areas pursuant to visions of such people as John Muir and Theodore Roosevelt. Not until the mid 20th century did efforts arise to target individual species for conservation, notably efforts in big cat conservation in South America led by the New York Zoological Society. By the 1970s, led primarily by work in the United States under the Endangered Species Act along with Biodiversity Action Plans developed in Australia, Sweden, the United Kingdom, hundreds of specific species protection plans ensued. The Society for Conservation Biology is a global community of conservation professionals dedicated to advancing the science and practice of conserving Earth's biological diversity.

Conservation of Biological Diversity

Biodiversity provides many ecosystem services that are often not readily visible. It plays an essential part in regulating the chemistry of our atmosphere, pollinating crops and generating water supply. Biodiversity is directly involved in recycling nutrients and providing fertile soils. Experiments with controlled environments have shown that humans cannot easily build ecosystems to support human needs; for example insect pollination cannot be mimicked by man-made construction. The total value of ecosystem services may amount to trillions of dollars in ecosystem services per annum to mankind. For example, one segment of North American forests has been assigned an annual value of 250 billion dollars; as another example, honey-bee pollination, a small segment of ecosystem services, is estimated to provide between 10 and 18 billion dollars of value per annum. The value of ecosystem services on one New Zealand island has been imputed to be as great as the GDP of that region.

Conservation biologists trace the ethics that guide their work back to early spiritual philosophies, including the Tao, Shinto, Hindu, Islamic and Buddhist traditions. In the West, origins of concern for the destruction of the natural environment by man can be traced to Plato; however, modern roots of conservation biology can be found in the late 18th century Enlightenment period particularly in England and Scotland. A number of thinkers, among them notably Lord Monboddo, described the importance of "preserving nature"; much of this early emphasis had its origins in Christian theology. By the early 1800s biogeography was ignited through efforts of Von Humboldt, DeCandolle, Lyell and Darwin; their efforts, while important in relating species to their environments, fell short of actual conservation.

The term *conservation* came into use in the late 19th century and referred to the management, mainly for economic reasons, of such natural resources as timber, fish, game, topsoil, pastureland, and minerals, and also to the preservation of forests, wildlife, parkland, wilderness, and watersheds. Western Europe was the source of much 19th century progress for conservation biology, particularly the British Empire; however, the United States began making sizable contributions to this field starting with thinking of Thoreau and taking form in the United States Congress passing the Forest Act of 1891, John Muir's work and the founding of the Sierra Club in 1895, founding of the New York Zoological Society in 1895 and establishment of a series of national forests and preserves by Theodore Roosevelt from 1901 to 1909.

By the early 1970s national and international governmental agencies became more active in the conservation of biodiversity. Notably the United Nations acted to conserve sites of outstanding cultural or natural importance to the common heritage of mankind. The programme was adopted by the General Conference of UNESCO in 1972. As of 2006, a total of 830 sites are listed: 644 cultural, 162 natural. The first country to pursue aggressive biological conservation through national legislation was the USA, which passed back to back legislation in the Endangered Species Act and National Environmental Policy Act, which together injected major funding and protection measures to large scale habitat protection and threatened species research.

In 1980 a significant development was the emergence of the urban conservation movement. A local organisation was established in Birmingham, UK, a development followed in rapid succession in cities across the UK, then overseas. Although perceived as a grass-roots movement, its early development was driven by a burgeoning of academic research into urban wildlife. Initially perceived as radical, the urban movement's view of conservation being inextricably linked with other human activity has now become mainstream in conservation thought. Considerable research effort is now directed at urban conservation biology.

By 1992 most of the countries of the world had become committed to the principles of conservation of biological diversity with the Convention on Biological Diversity; subsequently many countries began programmes of Biodiversity Action Plans to identify and conserve threatened species within their borders, as well as protect associated habitats.

Since 2000 the concept of landscape scale conservation has risen to prominence, with less emphasis being given to single-species or even single-habitat focused actions. Instead an ecosystem approach is advocated by most mainstream conservationist, although concerns have been expressed by those working to protect some high-profile species.

The science of ecology has clarified the workings of the biosphere; i.e., the complex interrelationships among humans, other species, and the physical environment; moreover, the burgeoning human population, and associated agriculture, industry and its ensuing pollution have demonstrated how easily ecological relationships can be disrupted.

India is the seventh largest country in the world and Asia's second largest nation with an area of 3,287,263 square km. The Indian mainland stretches from 8 4' to 37 6' N latitude and from 68 7' to 97 25' E longitude. It has a land frontier of some 15,200 kms and a coastline of 7,516 km. India's northern frontiers are with Xizang (Tibet) in the Peoples Republic of China, Nepal and Bhutan. In the north-west, India borders on Pakistan; in the north-east, China and Burma; and in the east, Burma.

The southern peninsula extends into the tropical waters of the Indian Ocean with the Bay of Bengal lying to the south-east and the Arabian Sea to the south-west. For administrative purposes India is divided into 24 states and 7 union territories. The country is home to around 846 million people, about 16% of the World's population. The climate of India is dominated by the Asiatic monsoon, most importantly by rains from the south-west between June and October, and drier winds from the north between December and February. From March to May the climate is dry and hot.

Wetlands

India has a rich variety of wetland habitats. The total area of wetlands (excluding rivers) in India is 58,286,000ha, or 18.4% of the country, 70% of which comprises areas under paddy cultivation. A total of 1,193 wetlands, covering an area of about 3,904,543 ha, were recorded in a preliminary inventory coordinated by the Department of Science and Technology, of which 572 were natural. Two sites - Chilka Lake (Orissa) and Keoladeo National Park (Bharatpur) - have been designated under the Convention of Wetlands of International Importance as being especially significant waterfowl habitats. The country's wetlands are generally differentiated by region into eight categories: the reservoirs of the Deccan Plateau in the south, together with the lagoons and the other wetlands of the southern west coast; the vast saline expanses of Rajasthan, Gujarat and the gulf of Kachchh; freshwater lakes and reservoirs from Gujarat eastwards through Rajasthan (Kaeoladeo Ghana National park) and Madhya Pradesh; the delta wetlands and lagoons of India's east coast (Chilka Lake); the freshwater marshes of the Gangetic Plain; the floodplain of the Brahmaputra; the marshes and swamps in the hills of north-east India and the Himalayan foothills; the lakes and rivers of the montane region of Kashmir and Ladakh; and the mangroves and other wetlands of the island arcs of the Andamans and Nicobars.

Forests

India possesses a distinct identity, not only because of its geography, history and culture but also because of the great diversity of its natural ecosystems.The panorama of Indian forests ranges from evergreen tropical rain forests in the Andaman and Nicobar Islands, the Western Ghats, and the north-eastern states, to dry alpine scrub high in the Himalaya to the north. Between the two extremes, the country has semi-evergreen rain forests, deciduous monsoon forests, thorn forests, subtropical pine forests in the lower montane zone and temperate montane forests.

One of the most important tropical forests classifications was developed for Greater India and later republished for present-day India. This approach has proved to have wide application outside India. In it 16 major forests types are recognised, subdivided into 221 minor types. Structure, physiognomy and floristics are all used as characters to define the types.

The main areas of tropical forest are found in the Andaman and Nicobar Islands; the Western Ghats, which fringe the Arabian Sea coastline of peninsular India; and the greater Assam region in the north-east. Small remnants of rain forest are found in Orissa state. Semi-evergreen rain forest is more extensive than the evergreen formation partly because evergreen forests tend to degrade to semi-evergreen with human interference. There are substantial differences in both the flora and fauna between the three major rain forest regions.

The Western Ghats Monsoon forests occur both on the western (coastal) margins of the ghats and on the eastern side where there is less rainfall. The distribution of forest in Kerala State, which contains part of the Western Ghats range. These forests contain several tree species of great commercial significance (e.g. Indian rosewood *Dalbergia latifolia*, Malabar Kino *Pterocarpus marsupium*, teak and *Terminalia crenulata*), but they have now been cleared from many areas.

In the rain forests there is an enormous number of tree species. At least 60 percent of the trees of the upper canopy are of species which individually contribute not more than one percent of the total number. Clumps of bamboo occur along streams or in poorly drained hollows throughout the evergreen and semi-evergreen forests of south-west India, probably in areas once cleared for shifting agriculture.

The tropical vegetation of north-east India typically occurs at elevations up to 900 m. It embraces evergreen and semi-evergreen rain forests, moist deciduous monsoon forests, riparian forests, swamps and grasslands. Evergreen rain forests are found in the Assam Valley, the foothills of the eastern Himalayas and the lower parts of the Naga Hills, Meghalaya, Mizoram, and Manipur where the rain fall exceeds 2300 mm per annum. In the Assam Valley the giant *Dipterocarpus macrocarpus* and *Shorea assamica* occur singly, occasionally attaining a girth of up to 7 m and a height of up to 50 m. The monsoon forests are mainly moist sal *Shorea robusta* forests, which occur widely in this region.

The Andamans and Nicobar islands have tropical evergreen rain forests and tropical semi-evergreen rainforests as well as tropical monsoon moist monsoon forests. The tropical evergreen rain forest is only slightly less grand in stature and rich in species than on the mainland. The dominant species is *Dipterocarpus grandiflorus* in hilly areas, while *Dipterocarpus kerrii* is dominant on some islands in the southern parts of the archipelago. The monsoon forests of the Andamans are dominated by *Pterocarpus dalbergioides* and *Terminalia* spp.

Marine Environment

The nearshore coastal waters of India are extremely rich fishing grounds. The total commercial marine catch for India has stabilised over the last ten years at between 1.4 and 1.6 million tonnes, with fishes from the clupeoid group (e.g. sardines *Sardinella* sp., Indian shad *Hilsa* sp. and whitebait *Stolephorus* sp.) accounting for approximately 30% of all landings. In 1981 it was estimated that there were approximately 180,000 non-mechanised boats (about 90% of India's fishing fleet) carrying out small-scale, subsistence fishing activities in these waters. At the same time there were about 20,000 mechanised boats and 75 deep-sea fishing vessels operating mainly out of ports in the states of Maharashtra, Kerala, Gujarat, Tamil Nadu and Karnataka.

Coral reefs occur along only a few sections of the mainland, principally the Gulf of Kutch, off the southern mainland coast, and around a number of islands opposite Sri Lanka. This general absence is due largely to the presence of major river systems and the sedimentary regime on the continental shelf. Elsewhere, corals are also found in Andaman, Nicobar, and Lakshadweep island groups although their diversity is reported to be lower than in south-east India.

Indian coral reefs have a wide range of resources which are of commercial value. Exploitation of corals, coral debris and coral sands is widespread on the Gulf of Mannar and Gulf of Kutch reefs, while ornamental shells, chanks and pearl oysters are the basis of an important reef industry in the south of India. Sea fans and seaweeds are exported for decorative purposes, and there is a spiny lobster fishing industry along the south-east coast, notably at Tuticorin, Madras and Mandapam Commercial exploitation of aquarium fishes from Indian coral reefs has gained importance only recently and as yet no organised effort has been made to exploit these resources. Reef fisheries are generally at the subsistence level and yields are unrecorded.

Other notable marine areas are seagrass beds, which although not directly exploited are valuable as habitats for commercially harvested species, particularly prawns, and mangrove stands. In the Gulf of Mannar the green tiger prawn *Penaeus semisulcatus* is extensively harvested for the export market. Seagrass beds are also important feeding areas for the dugong *Dugong dugon*, plus several species of marine turtle.

Five species of marine turtle occur in Indian waters: Green turtle *Chelonia mydas*, Loggerhead *Caretta caretta*, Olive Ridley *Lepidochelys olivacea*, Hawksbill *Eretmochelys imbricata* and Leatherback *Dermochelys coriacea*. Most of the marine turtle populations found in the Indian region are in decline. The principal reason for the decrease in numbers is deliberate human predation. Turtles are netted and speared along the entire Indian coast. In south-east India the annual catch is estimated at 4,000-5,000 animals, with *C. mydas* accounting for about 70% of the harvest. *C. caretta* and *L. olivacea* are the most widely consumed species. *E. imbricata* is occasionally eaten but it has caused deaths and so is usually caught for its shell alone. *D. coriacea* is boiled for its oil which is used for caulking boats and as protection from marine borers. Incidental netting is widespread. In the Gulf of Mannar turtles are still reasonably common near seagrass beds where shrimp trawlers operate, but off the coast of Bengal the growing number of mechanised fishing boats has had the effect of increasing incidental catch rates shows known turtle nesting areas in the Andaman Islands.

Species Diversity

India contains a great wealth of biological diversity in its forests, its wetlands and in its marine areas. This richness is shown in absolute numbers of species and the proportion they represent of the world total.

India has a great many scientific institutes and university departments interested in various aspects of biodiversity. A large number of scientists and technicians have been engaged in inventory, research, and monitoring. The general state of knowledge about the distribution and richness of the country's biological resources is therefore fairly good.

Inventories of birds, mammals, trees, fish and reptiles are moderately complete. Knowledge of special interest groups such as primates, pheasants, bovids, endemic birds, orchids, and so on,is steadily improving through collaboration of domestic scientists with those from overseas. The importance of these biological resources cannot be overestimated for the continued welfare of India's population.

Endemic Species

India has many endemic plant and vertebrate species. Among plants, species endemism is estimated at 33% with c. 140 endemic genera but no endemic families. Areas rich in endemism are north-east India, the Western Ghats and the north-western and eastern Himalayas. A small pocket of local endemism also occurs in the Eastern Ghats. The Gangetic plains are generally poor in endemics, while the Andaman and Nicobar Islands contribute at least 220 species to the endemic flora of India.

WCMC's Threatened Plants Unit (TPU) is in the preliminary stages of cataloguing the world's centres of plant diversity; approximately 150 botanical sites worldwide are so far recognised as important for conservation action, but others are constantly being identified. Five locations have so far been issued for India: the Agastyamalai Hills, Silent Valley and New Amarambalam Reserve and Periyar National Park, and the Eastern and Western Himalaya.

Endemism among mammals and birds is relatively low. Only 44 species of Indian mammal have a range that is confined entirely to within Indian territorial limits. Four endemic species of conservation significance occur in the Western Ghats. They are the Lion-tailed macaque *Macaca silenus*, Nilgiri leaf monkey *Trachypithecus johni*, Brown palm civet *Paradoxurus jerdoni* and Nilgiri tahr *Hemitragus hylocrius*.

Only 55 bird species are endemic to India, with distributions concentrated in areas of high rainfall. They are located mainly in eastern India along the mountain chains where the monsoon shadow occurs, south-west India (the Western Ghats), and the Nicobar and Andaman Islands.

In contrast, endemism in the Indian reptilian and amphibian fauna is high. There are around 187 endemic reptiles, and 110 endemic amphibian species. Eight amphibian genera are not found outside India. They include, among the caecilians, *Indotyphlus*, *Gegeneophis* and *Uraeotyphlus*; and among the anurans, the toad *Bufoides*, the microhylid *Melanobatrachus*, and the frogs *Ranixalus*, *Nannobatrachus* and *Nyctibatrachus*. Perhaps most notable among the endemic amphibian genera is the monotypic *Melanobatrachus* which has a single species known only from a few specimens collected in the Anaimalai Hills in the 1870s. It is possibly most closely related to two relict genera found in the mountains of eastern Tanzania.

Threatened Species

India contains 172 species of animal considered globally threatened by IUCN, or 2.9% of the world's total number of threatened species. These include 53 species of mammal, 69 birds, 23 reptiles and 3 amphibians. A full list of these species is given in Appendix 5. India contains globally important populations of some of Asia's rarest animals, such as the Bengal Fox, Asiatic Cheetah, Marbled Cat, Asiatic Lion, Indian Elephant, Asiatic Wild Ass, Indian Rhinoceros, Markhor, Gaur, Wild Asiatic Water Buffalo etc.

Area of Forest Land

Some 50 million hectares, about 17 percent of India's land area, were regarded as forest land in the early 1990s. In FY 1987, however, actual forest cover was 64 million hectares. However, because more than 50 percent of this land was barren or brushland, the area under productive forest was actually less than 35 million hectares, or approximately 10 percent of the country's land area. The growing population's high demand for forest resources continued the destruction and degradation of forests through the 1980s, taking a heavy toll on the soil. An estimated 6 billion tons of topsoil were lost annually. However, India's 0.6 percent average annual rate of deforestation for agricultural and nonlumbering land uses in the decade beginning in 1981 was one of the lowest in the world and on a par with Brazil.

Many Indian forests in the mid-1990s are found in high-rainfall, high-altitude regions, areas to which access is difficult. About 20 percent of total forestland is in Madhya Pradesh; other states with significant forests are Orissa, Maharashtra, and Andhra Pradesh (each with about 9 percent of the national total); Arunachal Pradesh (7 percent); and Uttar Pradesh (6 percent). The variety of forest vegetation is large: there are 600 species of hardwoods, *sal* (*Shorea robusta*) and teak being the principal economic species.

Conservation has been an avowed goal of Indian government policy since Indian independence. Afforestation increased from a negligible amount in the first plan to nearly 8.9 million hectares in the seventh plan. The cumulative area afforested during the 1951-91 period was nearly 17.9 million hectares. However, despite large-scale tree planting programs, forestry is one arena in which India has actually regressed since independence. Annual fellings at about four times the growth rate are a major cause. Widespread pilfering by villagers for firewood and fodder also represents a major decrement. In addition, the forested area has been shrinking as a result of land cleared for farming, inundations for irrigation and hydroelectric power projects, and construction of new urban areas, industrial plants, roads, power lines, and schools.

India's long-term strategy for forestry development reflects three major objectives: to reduce soil erosion and flooding; to supply the growing needs of the domestic wood products industries; and to supply the needs of the rural population for fuelwood, fodder, small timber, and miscellaneous forest produce. To achieve these objectives, the National Commission on Agriculture in 1976 recommended the reorganisation of state forestry departments and advocated the concept of social forestry. The commission itself worked on the first two objectives, emphasizing traditional forestry and wildlife activities; in pursuit of the third objective, the commission recommended the establishment of a new kind of unit to develop community forests. Following the leads of Gujarat and Uttar Pradesh, a number of other states also established community-based forestry agencies that emphasized programs on farm forestry, timber management, extension forestry, reforestation of degraded forests, and use of forests for recreational purposes.

Such socially responsible forestry was encouraged by state community forestry agencies. They emphasized such projects as planting wood lots on denuded communal cattle-grazing grounds to make villages self-sufficient in fuelwood, to supply timber needed for the construction of village houses, and to provide the wood needed for the repair of farm implements. Both individual farmers and tribal communities were also encouraged to grow trees for profit. For example, in Gujarat, one of the more aggressive states in developing programs of socioeconomic importance, the forestry department distributed 200 million tree seedlings in 1983. The fast-growing eucalyptus is the main species being planted nationwide, followed by pine and poplar.

The role of India's forests in the national economy and in ecology was further emphasized in the 1988 National Forest Policy, which focused on ensuring environmental stability, restoring the ecological balance, and preserving the remaining forests. Other objectives of the policy were meeting the need for fuel wood, fodder, and small timber for rural and tribal people while recognising the need to actively involve local people in the management of forest resources. Also in 1988, the Forest Conservation Act of 1980 was amended to facilitate stricter conservation measures. A new target was to increase the forest cover to 33 percent of India's land area from the then-official estimate of 23 percent. In June 1990, the central government adopted resolutions that combined forest science with social forestry, that is, taking the sociocultural traditions of the local people into consideration.

Since the early 1970s, as they realised that deforestation threatened not only the ecology but their livelihood in a variety of ways, people have become more interested and involved in conservation. The best known popular activist movement is the Chipko Movement In India, in which local women decided to fight the government and the vested interests to save trees. The women of Chamoli District, Uttar Pradesh, declared that they would embrace—literally "to stick to" (*chipkna* in Hindi)—trees if a sporting goods manufacturer attempted to cut down ash trees in their district. Since initial activism in 1973, the movement has spread and become an ecological movement leading to similar actions in other forest areas. The movement has slowed down the process of deforestation, exposed vested interests, increased ecological awareness, and demonstrated the viability of people power

Reserved forests and protected forests

A reserved forest (also called reserve forest) or a protected forest in India are terms denoting forests accorded a certain degree of protection. The terms were first introduced in the Indian Forest Act, 1927 in British India, to refer to certain forests granted protection under the British crown in British India, but not associated suzerainties. After Indian independence, the Government of India retained the status of the existing reserved and protected forests, as well as incorporating new reserved and protected forests. A large number of forests which came under the jurisdiction of the Government of India during the political integration of India were initially granted such protection.

Land rights to forests declared to be Reserved forests or Protected forests are typically acquired (if not already owned) and owned by the Government of India. Unlike national parks of India or wildlife sanctuaries of India, reserved forests and protected forests are declared by the respective state governments. As of present, reserved forests and protected forests differ in one important way:

— Rights to all activities like hunting, grazing, etc in *reserved forests* are banned unless specific orders are issued otherwise

— Rights to all activities like hunting, grazing, etc in *protected forests* are allowed unless specific orders are issues otherwise

In reserved forests, rights to activities like hunting and grazing are sometimes given to communities living on the fringes of the forest, who sustain their livelihood partially or wholly from forest resources or products. Thus, typically reserved forests enjoy a higher degree of protection with respect to protected forests. However, it is possible that certain protected forests may enjoy more protection with respect to certain reserved forests.

Protected forests are of two kinds - demarcated protected forests and undemarcated protected forests, based on whether the limits of the forest have been specified by a formal notification.

Typically, reserved forests are often upgraded to the status of wildlife sanctuaries, which in turn may be upgraded to the status of national parks, with each category receiving a higher degree of protection and government funding. For example, Sariska National Park was declared a reserved forest in 1955, upgraded to the status of a wildlife sanctuary in 1958, becoming a Tiger Reserve in 1978. Sariska became a national park in 1992, though primary notification to declare it as a national park was issued as early as 1982.

At a time when conservation of the earth's biological diversity (bio-diversity) has become more of a social challenge than ecological science, several initiatives have arisen worldwide. And although this transition widely prevailed since the 1972 Stockholm Conference on Human and Environment Development, the UN Convention on Biological Diversity, that came into force in 1993, fully reinforced the need for a social approach to conservation and sustainable use of the earth's biodiversity.

The main emphasis of the UN Convention on Biological Diversity (CBD) is on conservation, sustainable use and equitable sharing of the benefits that arise from the use of biodiversity. All of the nearly 175 nations that are signatories to the CBD are thus obliged to nurture the biodiversity within their respective political boundaries guided by the provisions of the CBD.

India is a signatory of the CBD since February 1994. Amongst the signatory nations, India has some unique qualities. Firstly, it has been globally ranked amongst the 12 megadiversity countries. Two of its biogeographic provinces, viz. Western Ghats and Eastern Himalayas are considered as 'hot spots' of biodiversity. In fact there are only 18 such hot spots throughout the world. Despite India's large human population and the great diversity of cultures, its land and waters shelter a wide variety of natural habitats and an estimated 500,000 species of living organisms. It is also widely acknowledged that such a remarkable heritage of biodiversity is not merely the product of the varied geography of the country but its people and their time-tested conservation traditions.

Efforts to conserve India's biodiversity have constantly evolved over the millennium. However, what has often been projected as 'conservation' in the country is the more recent system of protected areas: wildlife sanctuaries, national parks and biosphere reserves. These are all government initiatives and despite being successful in declaring nearly 5% of the country's surface area as legally protected, the entire system has attracted a considerable amount of criticism and has often been rated as 'flawed' by the import of western ideals – ideals that exclude people from the parks.

Historically the main threat to biodiversity has been a set of threats generated from the overpopulation of humans: mass agriculture, deforestation, overgrazing, slash-and-burn, urban development, pesticide use. Worldwide, the effects of global warming add a potentially catastrophic threat to global biological diversity; a 2004 study by Chris Thomas, Lee Hannah, et al. estimated that 15 to 37 percent of all species would become extinct by 2050.

Planning for Environmental Conservation and Protection

Conventionally, the environmental pollution problems are solved by introducing environmental management techniques such as control of pollution at source, providing of sewage treatment facilities etc. However, environmental risks are not being controlled completely by such solutions. The environmental aspects are to be induced into each of the developmental activities at the planning stage itself and are to be well co-ordinated and balanced. Presently, the environmental aspects are not usually considered while preparing master plans or regional plans and the process is skewed towards developmental needs. For all developmental activities, a crucial input is land and depending on the activity a specific landuse is decided. The environmentally related landuse such as trade and industry, housing construction, mining etc. are likely to have some impact on the environment. These land uses need proper planning and integration as some of the activites have interdependencies auch as industry with tranpsort, housing etc.

The spatial planning tools can help in sustainable development. In India, presently spatial planning approach is mostly limited to urban areas only and the regions are not normally considered for planning purposes and for attaining balanced development. The country today lacks integrated spatial planning (national/state/regional/town level). The planning is mostly limited to urban areas and even in these areas the master plans do not taken into consideration the environmetnal aspects and the developmental needs are not well reflected. Also, the master plans are several times are violated. Lack of planning is leading to unbalanced development thereby forming uneconomical agglomerations, ecologically degraded areas and over exploitation of resources. The developmental activities tend to be haphazard and uncontrolled thus leading to over use, congestion, poor land use compatibility etc.

Challenges of Integrated Spatial Planning

The planning solutions for achieving balanced and sustainable development had been demonstrated to a good extent in some of the countries. Some of the major constraints for introducing integrated spatial planning in India are:

— In view of the existing social and living conditions, economic interests may tend to over-ride the environmental aspects;

— Ecosystem are already over-used in some areas;

— Introduction of spatial planning which involves highly complex nature of planning activities is a daunting task particularly in a large country, like India;

— Lack of legal framework for spatial planning, dearth of financial resources, inadequate environmental awareness, shortage of manpower and limitations in technical competence are among the constraints in integration of environmental concerns in the development process.

However, spatial planning based on assessment of existing environmental profiles as well as potential assimilative capacity could help environmentally acceptable development and resolve the conflicts which are otherwise confronted with. Planning of activities based on assessment of local or regional environmental impacts could be a useful approach for introducing the concept of spatial planning in a limited manner under Indian conditions.

Lack of Environmental Planning

Presently, the environmental aspects are not usually considered while preparing master plans and the process is skewed towards developmental needs. For all developmental activities, a crucial input is land and depending on the activity a specific landuse is decided. The environmentally relevant land uses are trade and commerce, housing construction, transport facilities (road, rail and water), utilities (water - surface and ground etc.), refuse/hazardous waste disposal facilities, wastewater installations, quarrying and mining, power generation, forestry, recreation and tourism etc. These land uses are likely to have impact on the environment. There is a need for assessment of the land in terms of not only the economic aspects but also the environmental aspects and the land uses are accordingly to be allocated so that the natural environment and ecological balance is not disturbed.

Inadequacy of Conventional Control Techniques

The environmental problems of concern and increased environmental risks are due to air pollution from vehicular, industrial and domestic sources, noise pollution, water pollution - lack of proper storm water drainage and sewerage system, improper and inadequate garbage collection and disposal system, haphazard siting of industries/processes, transportation, storage and handling of toxic or hazardous chemicals, lack of adequate open spaces and green areas; etc. Conventionally, the environmental pollution problems are solved by introducing environmental management techniques such as control of pollution at source, providing of sewage treatment facilities etc. These measures are proving to be inadequate because of the complexity associated with the dynamics of development.

Inadequacy of Conventional Control Techniques

The environmental problems of concern and increased environmental risks are due to air pollution from vehicular, industrial and domestic sources, noise pollution, water pollution - lack of proper storm water drainage and sewerage system, improper and inadequate garbage collection and disposal system, haphazard siting of industries/processes, transportation, storage and handling of toxic or hazardous chemicals, lack of adequate open spaces and green areas; etc. Conventionally, the environmental pollution problems are solved by introducing environmental management techniques such as control of pollution at source, providing of sewage treatment facilities etc. These measures are proving to be inadequate because of the complexity associated with the dynamics of development.

Increasing Public Awareness

There is an increase in public awareness on pollution and its affects. The people today are demanding good quality of life and living conditions. The increasing public interest litigation (PILs) for relocating environmentally incompatible land uses is an indication that there will be an increased need for proper planning of land uses and siting of industries and other development projects.

Growing Environmental Costs

It has been proved even with in our country that though the economic considerations tend to bring in gains in a short term, the liabilities from neglecting the environmental aspects are heavier in long run. The costs involved for cleaning up river Ganga or for introduction of unleaded petrol or for shifting industries from Delhi are just a few examples. This necessiates proper planning in advance so as to be prepared for the subsequent consequences.

Constraints in the Existing Industrial Siting Procedures

- The targets for industrial development are fixed but the sites for these industries to come up are rarely pre-determined thereby paving the way for haphazard siting of industries.
- The responsibility of selecting a site is primarily entrusted with the entrepreneurs and this does not necessarily lead to objective assessment of environmental aspects.
- The information base available for evaluating environmental impacts and taking decisions on industrial siting is weak. Hence, it causes subjectivity in decision- making process as well as lack of transparency and delay.

— The environmental clearance by the regulatory authorities does not necessarily imply zero pollution from an industry.

Hence, the major challenge is not just finding a site for an industry or a developmental activity but is finding a solution for achieving sustainable development. It is being increasingly realised that the developmental activities are to be planned in such a way that the socio-economic objectives are fulfilled without causing adverse impacts on the environment.

Possibilities

The possibilities for a suitable solution for the Indian conditions include introduction of integrated spatial planning as a long term solution. In the context of spatial planning, the planning models of other countries having similar conditions/constraints with respect to population, resources etc. can be taken as an example for working out suitable solutions for Indian conditions. For example, the German planning system in based on 'co-operation' among various levels - federal, state, regional, local etc. and 'balancing' among different sectors - industry, agriculture, forestry, environment etc. The prior interaction with the lower level makes the guidelines more acceptable and the plans more implementable on ground. At the same time, this helps achieve co-ordinated and balanced development.

For the situations in our country, sectoral land use plans for all the environmentally relevant activites such as those given below should be prepared keeping in view the developmental needs/targtes and the environmental considerations and then these are to be integrated into one plan that is binding on all:

— trade and industry locations;

— housing construction;

— transport facilities (road, rail, water)

— utilities;

— refuse/haz. waste and wastewater installations;

— quarrying/mining;

— power generation;

— agriculture;

— forestry;

— inland and coastal fisheries;

— recreation and tourism;
— water regulation and development;
— tapping of groundwater; and
— outfalls into surfacewater.

This helps individual sectoral authorities to meet their development targets while ensuring that these targets are achieved in an environmentally compatible manner.

Environmental Planning Initiatives

Environmental planning is a relatively new tool for environmental protection in India. Historically, the Central and the State Pollution Control Boards were entrusted with environmental protection with emphasis on control and abatement of industrial pollution.

The prevailing situation of industrial siting and incompatible surrounding land uses demands adoption of more reliable and long-lasting solutions. The need for environmental planning was understood by CPCB and the Ministry of Environment & Forests (MoEF), Govt. of India. Consequently, certain pilot studies were taken up at Central as well as State level. Experience with this type of studies, in particular in the Union Territory of Pondicherry and for Hassan District in Karnataka, stimulated CPCB and SPCBs to start a programme on developing necessary capacities for environmental planning within the environmental administration.

The provisions for this strategic development are founded in the Environment (Protection) Act, 1986, which authorises the Central Government "to take all such measures as it seems necessary for the purpose of protecting and improving the quality of the environment and preventing, controlling and abating environmental pollution" [Section 3(1)]. Measures under this clause may include "planning and execution of a nation-wide programme for the prevention, control and abatement of environmental pollution" [Section 3(2)], (ii)). This task of environmental management includes also spatial (geographical) aspects as explicitly mentioned under Section 3 (2) (v) "restriction of areas in which any industries, operation or processes shall not be carried out or shall be carried out subject to certain safeguards."

The work in the first phase of the programme started in early 1995 with the conduct of pilot studies on preparation of Zoning Atlas for Siting of Industries based on environmental considerations in selected 19 Districts of 14 States. Based on the response received, the programme had been expanded and intensified under the World bank funded Environmental Mangement Capacity Building Project. The goal for the programme has been formulated as follows:

> "Technical, instrumental and institutional capacities needed for producing spatial environmental assessments for planning purposes are established or strengthened in order to produce the targeted studies whose results could be used to effectively promote the environmentally compatible spatial planning in India"

This formulation reflects the thrust of the programme which not only includes capacity building and strengthening but in particular the use of the capacities built-up in the environmental administration of India to promote environmentally compatible spatial planning.

The purpose of the programme has been formulated as follows:

> "To strengthen and increasingly utilise competence, instruments and the institutional basis for environmentally compatible, sustainable management of land and land based natural resources, in order to harmonise spatial development and environment in India "

The activites under the programme have been intiated at national and State levels for preparing information base on environment and at the District level for zoning the areas for sititn of industries, at microlevel (1:50,000) for identification of sites for industrial estates and at the city level for preparation of enviroentnal mangement plans for improvement of environmental quality.

The programme is being well received and it is hoped that the initiatives of CPCB will go a long way in helping developmental objectives in an environmentally sound manner.

Recent initiatives of environmental conservation, particularly in tropical parts of the world, are unduly oriented towards biodiversity, ignoring ecosystem services, though conserving biodiversity is often justified on the ground that it contributes to ecosystem services . Recent exercises on the prioritisation of creatures and places deserving most attention for conservation, and on developing national biodiversity conservation plans in several Asian countries, may further increase this imbalance. These also include India's National Biodiversity Strategy and Action Plan (NBSAP), regarded as one of the greatest exercises on conservation ever taken place in the developing world.

Because conservation is tightly connected with funding, attempts are being made to find ways to reduce costs and maximise benefits of biodiversity conservation. For example, Myers *et al.* argue that by protecting 25 top hot spots, comprising only 1.4% of the earth's surface, 44% of all plant species and 35% of vertebrate species worldwide can be saved. It is important to develop strategies to conserve as high a proportion of the species on the planet as possible, but a balanced approach of conservation should also consider other environmental issues such as carbon sequestration, waste dissipation, the hydrological balance, soil formation, health of local ecosystems, and others not intimately connected with biodiversity.

Ecosystem Services

Ecosystem services are defined as services generated due to the interaction and exchange between biotic and abiotic components of an ecosystem (Figure 1). Accordingly, ecosystem services do not include ecosystem products such as food. Ecosystem services are generated by ecosystem functions (such as production and nutrient cycling), but the functions and services do not necessarily show a one-to-one correspondence. There are numerous ways through which interactions among the community, energy flow and cycling of materials generate services to humans.

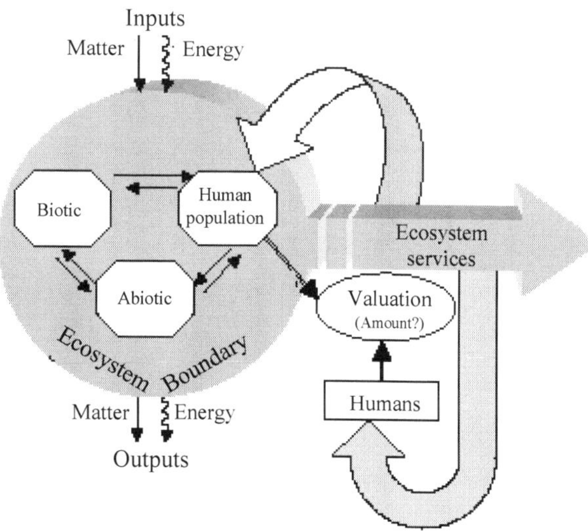

Figure 1: Representation of connections between the components of a circumscribed ecosystem, and between ecosystem services and humans.

The ecosystem services commonly listed are carbon sequestration, purification of water and air, soil formation and renewal of fertility, regulation of water and nutrient movement from one ecosystem to another, dissipation of wastes and removal of toxins, control over climate, dampening of fluctuations in physical factors of the environment, and providing recreation and scenic beauty. Since ecosystem services are considered in the context of humans (Figure 1), their importance would largely depend on the size of the population that utilises them.

Limitations of Biodiversity-centred Approach

The biodiversity-centred approach to conserving nature has some serious limitations. It is primarily based on giving the highest priority to areas of highest diversity, and identifying hot spots and other species-rich areas, generally located in the tropics. Many rare and restricted species occur outside species-rich areas, and it is not necessary that areas rich in one kind of organism are also rich in others. This approach neglects the areas not so rich in biodiversity, but important from the standpoint of the ecosystem services they generate for humans. To cite an Indian example, the Western Himalayan Ecoregion (WHE) in India is not as species-rich as the Eastern Himalayan Ecoregion (EHE), and the latter along with the Western Ghats is among the major hot spots of the world. However, the importance of WHE as the provider of ecosystem services is clearly greater than that of the two Indian hot spots. Associated with the WHE is the Gangetic plain, easily one of the most productive and robust agricultural zones on the planet, supporting nearly 400 million people.

Though the Gangetic plain owes its origin and development largely to geological processes, the nursing effect of the forests of WHE, in terms of regulated supply of soil fertility and water, has played an important role in continued food production from the region for at least 6000 years. An ecosystem subsidy such as this was recognised long ago by Odum as accounting for the high productivity of estuaries. Also, the alpine meadows of WHE, spread over about 70,000 km^2, may prove to be critical for the woody species of lower altitudes, that would be forced to migrate upward in the event of global warming. There are reasons to conserve refugia that may be important in the changing future, irrespective of the amount of diversity they have, particularly in view of the fact that origins of biodiversity are poorly understood. If all values were considered, both WHE and EHE should get attention from conservation planners, but if a conservation plan is centred around biodiversity alone, WHE may be neglected.

At least certain ecosystem services are not related to biodiversity. For example, boreal peat lands are among the most speciespoor ecosystems, but storing about 23–26% of all terrestrial organic carbon they play a significant role in the control of regional and global climate through internal ecosystem control of energy fluxes.

It is important to create and align diverse incentives for conservation, particularly because opportunity costs of conservation, especially of biodiversity, are perceived to be too high. In poor countries, it is difficult to motivate people to protect species by arguing that useful products from them await discovery. In contrast, ecosystem services are already in use; they need only to be identified and to have people educated about them.

The limitations of the biodiversity-oriented approach to the conservation of nature in poor countries become particularly apparent outside protected areas, where an economic rationale is generally sought for forest conservation. The basis of a market-oriented conservation is that the extraction of useful species from the natural ecosystem is sustainable and economically profitable. Such requirements would focus attention only on ecosystems in which economically important species occur in high densities. The concentration of biomass in a few economically useful species means the promotion of a system with low biodiversity. Therefore, areas of high biodiversity will require non-market mechanisms to ensure their protection, unless values of ecosystem services are connected with market mechanism. Thus, even to justify the conservation of biodiversity-rich areas economically, there may be a need to take the support of ecosystem services as incentives. Valuation of ecosystem services improves the economics of both species-rich and speciespoor ecosystems.

It may be needed to expand the list of components of conservation further. For example, in a watershed there may be a need to conserve geological material also, that has little biota but filters or/and stores water for drinking. Porous limestone rocks and debris deposited at the bottom of a catchment are good examples of natural water filters. Such watershed components also play a significant role in regional biogeochemical cycles, in linking ecosystems, and generating services in the context of a greater spatial scale. The New York City water department decided to allocate $ 1.5 billion to preserve the Catskill and Delware watersheds, rather than spending $ 6 billion to construct a new water-treatment system. The high valuation is not because of high diversity of vegetation in the watershed, but because of the other ecosystem properties. In central Himalaya, a valley-fill located in the watershed of Lake Nainital is shown to contribute significantly to its health by providing filtered water, and diluting pollutants and controlling siltation.

An ecosystem approach is more effective even for protecting biodiversity. Developing economic enterprises based on the sustainable use of biodiversity is suggested for the conservation of biodiversity. However, the 50-year history of fisheries with the paradigm of sustainability of a single species is associated with numerous and repeated collapses of fish populations and alteration of marine ecosystems irreversibly. Pitcher suggests that to conserve fish species the goal of the management should be rebuilding the marine ecosystems, including the creations of no-take marine reserves. The conservation strategy that focuses on biodiversity-rich areas, though it may ensure protection of a high proportion of global biodiversity, ignores large areas low in diversity such as arid regions and high mountains.

However, it is essential to conserve the ecosystems of such areas (i) because of the day-to-day dependence of the people on local ecosystem services, and (ii) because they maintain a wide range of variation in life. It is important to recognise and conserve the ecosystems of extreme habitats, as such areas extend the global gradients of environment, communities and ecosystems.

The question of whether diversity improves ecosystem services is still unresolved. A number of experiments based on terrestrial conditions, generally carried out during the last seven years, suggest a direct relationship between species diversity and ecosystem functioning, measured variously, but the debate continues as strong doubts have been raised about the experimental designs. Any attempt to seek a relationship between species diversity and ecosystem services appears to be futile in the case of freshwaters, where the medium, water, is the principal determinant of ecosystem functioning. Here, ecosystem functioning is not at all synonymous with the absolute preservation of all species. In fact, many freshwater ecosystem services decline with the increase in species diversity.

For example, for recreation purposes, a species-poor oligotrophic lake has far more value than a species-rich eutrophic lake. There are several examples of comparative studies indicating a negative correlation between diversity and phytoplankton biomass or productivity in lakes. In freshwaters, the whole system must be considered from the point of view of conserving a particular species. Aquatic ecosystems in general are evaluated higher than terrestrial ecosystems for their services to humans, though some of the latter type may be exceptionally rich in biodiversity.

Expanding the Ecosystem Services

Most studies describe ecosystem services of a general nature, attributable to any ecosystem. For example, a forest is associated with services such as landscape and watershed stabilisation, soil protection, water and nutrient retention within the soil or the entire ecosystem, and regulation of water flows, thereby preventing flood-anddrought regimes in downstream territory, returning water from the ground into the atmosphere, buffering against the spread of pests and diseases, modulating climate through regulation of rainfall regimes and albedo, and reducing global warming by sequestering carbon and controlling dry-land salinity.

However, to develop a conservation plan for a region and to harness the ecosystem services sustainably, we need to realise that even the ecosystems of an area developed under similar climates show a wide range of attributes, depending on the species composition and plant functional type, the developmental stage of the ecosystem, connections with other ecosystems and the spatial pattern. There is a need to go beyond the stage of generalised ecosystem characters, and (i) consider services that emanate from a stand or other units of a particular forest type, (ii) determine how ecosystems of a region or landscape vary in their attributes, and how such variations affect the quality and quantity of services that humans receive and can harness, (iii) identify linkages across ecosystems, and their effects on ecosystem services in a regional context, (iv) identify and map areas generating ecosystem services and people consuming them or taking advantage of them, and (v) undertake economic valuation of ecosystem services and develop a suitable payment mechanism to assure their continued availability.

To cite a few examples, an alder (*Alnus*) forest is particularly useful for supplying nitrogen to plantations or croplands located downstream. In the Western Himalaya, an oak (*Quercus leucotrichophora*) forest serves most effectively in terms of soil development, protection of nutrients, water retention, and the life of connected springs in a watershed, while pine (*Pinus roxburghii*) conserves nutrients efficiently on steep slopes by employing a high proportional retranslocation of nutrients from senescing leaves, slow litter decomposition and microbial immobilisation of nutrients.

Studies on the relationship between hydrology, nutrient supply and plants show that they can affect the spatial pattern of services significantly even in simpler tundra ecosystems. The relationships between recharge and discharge areas, and forest functioning and services in terms of nutrient supply are not adequately described, with a few exceptions, even for forests occurring on an easily identifiable topographic gradient. Discharge areas in Fennoscandian boreal forests, in which low N and low pH are limiting to productivity, play an important role in providing a relatively stable environment and lead to the formation of productive communities, requiring high soil pH and nutrients. Attributes such as these can be harnessed profitably.

Though soil formation along with nutrient cycling is considered to be one of the premier ecosystem services, almost no information is available on how the rate of soil formation varies across different terrestrial ecosystems of a region under the influence of the same climate, or how the rate of soil formation is affected by subsequent deforestation. Forests may vary in numerous other features, such as the retrieval of nutrients from deeper soil layers, influence on hydrology and fire regimes, which may be suitably harnessed to generate services. Services of these ecosystems can be seen at various scales – local, regional, inter-regional and global.

A study on a northern hardwood forest indicates how a complex interplay among vegetation, topography, soil factors and soil microbes determines the characteristics of N-cycling. The form of fungi and their ability to form a mat in the forest soil and to release organic acids, affect the rate of mineral weathering of forest soils, enhancing mobility of important nutrients and their uptake by trees and other plants. Griffiths *et al*. showed that in Douglas-fir forest (*Pseudotsuga menziesii*) in the Pacific Northwest of USA, the concentrations of soluble phosphates and sulphates were significantly higher in fungal mats than in non-mat soils.

A study based on more than 2700 soil profiles indicated how ecosystems vary widely in distribution of carbon with soil depth and carbon sequestration ability because of differences in only two or three parameters such as plant functional type, allocation pattern to below-ground plant parts, and clay content in soil. Ecosystems with deeper plant roots and more clay particles in soil (through protecting and stabilising organic matter) are more effective as a long-term carbon sink than those with shallower roots. Soil organic carbon is important in relation to climatic change, as on a global scale soil stores 2344 Pg C, with 842 Pg C distributed in deeper soils (1–3 m depth) alone, which is more than carbon in the global atmosphere. An effective ecosystem pattern can be determined for a region to ameliorate rising atmospheric CO_2 levels.

To enable these services of great intrinsic value to command price calls for a radical change in thinking. As described earlier, the WHE of India is speculated to have a high ecosystem service value, partly because of its connection with a large territory down below. There is no such receiver territory of services in the case of EHE. In the case of the Western Himalayan croplands, the maintenance of genetic diversity of crops and fodder plants, soil microbes, and organically produced food grain or pulses can also be recognised as ecosystem services. The production of food grain and pulses largely depends on services generated by the adjoining forests.

There is a need to work out details of ecosystem services. For example, though vegetation is said to affect soil formation, how ecosystems of an area differ in this regard is hardly known. Some processes such as nitrogen fixation are considered to be a character of a species, but Nfixation occurs within the ecosystem arena, involving the interaction of the photosynthesis of host species, bacterial population, shade cast by plants, soil nutrients and soil moisture. Retention of nutrients and water within an area, involving contributions of mycorrhizae, tree growth, retranslocation of nutrients from senescing leaves, microbial immobilisation of nutrients and a host of other processes, is also a form of ecosystem service. For example, the age-old practice of shifting agriculture in the tropics depends on soil fertility developed by forests over several years.

A given ecosystem attribute can be put to various uses. For example, the property of decomposition of a natural area such as forest stand could be used to take care of orange peels by a juice company or for preparing manure from forest litter, as is done in the Himalaya. It is not easy to document all services precisely, and to make people perceive their roles. Equally difficult is the identification of providers and users of ecosystem services. We need data to analyse these, educate people and make them perceive the importance of ecosystem services. In democratic societies, it is the perception of individuals that determines the valuation of services provided, and individuals' preferences depend on how much they know about the environment and its significance to their lives. For example, people in the Himalaya would prefer to have forests if they knew enough about the total balance sheet of costs and benefits. The balance sheet may include, for example, productivity of different systems, such as forest and cropland, and services supplied from forest to maintain fertility of cropland, its role in crop yield, soil water storage, and carbon sequestration.

International Legal Regimes for Environmental Protection

International Environmental Law

International environmental law is the body of international law that concerns the protection of the global environment.

Originally associated with the principle that states must not permit the use of their territory in such a way as to injure the territory of other states, international environmental law has since been expanded by a plethora of legally-binding international agreements. These encompass a wide variety of issue-areas, from terrestrial, marine and atmospheric pollution through to wildlife and biodiversity protection.

The key constitutional moments in the development of international environmental law are:

— the 1972 United Nations Convention on the Human Environment (UNCHE), held in Stockholm, Sweden;

— the 1987 Brundtland Report, Our Common Future, which coined the phrase 'sustainable development';

— the 1992 United Nations Conference on Environment and Development (UNCED), held in Rio de Janeiro, Brazil

The 1972 United Nations Conference on the Human Environment focused on the 'human' environment. The conference issued the Declaration on the Human Environment, a statement containing 26 principles and 109 recommendations (now referred to as the Stockholm Declaration). The creation of an environmental agency was also approved, now known as UNEP. In addition, there was the adoption of a Stockholm Action Program. There were no legally binding outcomes resulting from the Stockholm Conference. Principle 21 of the Declaration was a restatement of law already in existence since Roman times, namely that of 'good neighbourliness'. The Action Plan was never successfully followed by any country.

The 1992 Rio conference (also known as the Earth Summit) led to the adoption several important legally binding environmental treaties, namely the 1992 United Nations Framework Convention on Climate Change and the 1992 Convention on Biological Diversity. In addition. the parties adopted a soft law Declaration on Environment and Development which reaffirmed the Stockholm Declaration and provided 27 principles guiding environment and development (now referred to as the Rio Declaration). Another influential soft law document that the parties adopted was Agenda 21, a guide to implementation of the treaties agreed to at the Summit and a guide as to the principles of sustainable development. Agenda 21 also established the United Nations Commission on Sustainable Development (CSD) and the Global Environment Facility (GEF). Finally, the non-legal, non-binding Forest Principles were formed at the Earth Summit.

A further meeting was held in 2002, known as the World Summit on Sustainable Development (WSSD), held in Johannesburg, South Africa. Notable is the absence from its title of the word 'environment'. Although this meeting was held to mark the tenth anniversary of the Earth Summit, it is considered by many environmentalists and environmental lawyers to have been less than successful in environmental terms. It attained only limited progress towards stricter global regulation of human impacts on the natural environment. Nonetheless the WSSD brought a renewed emphasis on the synergies between combatting poverty and improving the environment.

Sources of International Environmental Law

International environmental law derives its content from four main sources:

— International agreements (also called treaties, conventions, international legal instruments, pacts, protocols, covenants)

— Customary international law

— General principles of law

— Other/ new sources (e.g., court decisions (case-law), resolutions, declarations, doctrine, recommendations given by world organisations etc.)

International agreements

International environmental agreements can be bilateral, regional or multilateral in nature. The multilateral environmental agreements are frequently referred to as MEAs for short and have become far more common in recent decades. Treaty law is known as a traditional source of law.The majority of the conventions relating to international environmental law are specific; that means that they deal directly with environmental issues. There are some general treaties with one or two clauses referring to environmental issues but these are rarer. There are about 1000 environmental law treaties in existence today; no other area of law has generated such a large body of conventions on a specific topic.

Protocols

Protocols are like mini-agreements that "hang off" the main treaty. They exist in many areas of international law but are especially useful in the environment field, where they can be used to update scientific knowledge. They also permit countries to reach agreement on a framework agreement that would otherwise be contentious, by allowing the details to be left to a later date for determination. Protocols are generally much easier to generate than a treaty and they can enter into force very quickly. The most widely-known protocol in international environmental law is the Kyoto Protocol.

Customary international law

Customary international law is important in international environmental law. These are the norms and rules that countries follow as a matter of custom and they are so prevalent that they bind all states in the world. When a principle becomes customary law is not clear cut and many arguments are put forward by states not wishing to be bound.

Examples of customary international law relevant to the environment include:

— the duty to warn other states promptly about emergencies of an environmental nature and environmental damages to which another state or states may be exposed

— Principle 21 of the Stockholm Declaration ('good neighbourliness' or sic utere)

Judicial decisions

International environmental law also includes the opinions of international courts and tribunals. While there are few and they have limited authority, the decisions carry much weight with legal commentators and are quite influential on the development of international environmental law.

The courts include: the International Court of Justice (ICJ); the Law of the Sea Court; the European Court of Justice; regional treaty tribunals. Arguably the World Trade Organisation's Dispute Settlement Board (DSB) is getting a say on environmental law also.

Organising Principles

International environmental law is heavily influenced by a collection of organising principles.

As with international law, the chief guiding principle is that of sovereignty, which means that a country (state) has full power in its own territory to do as it pleases (subject to international laws it has agreed to). All other international environmental law principles evolved with this principle in the background and to varying degrees have either supported it or modified it to some extent.

Some of the organising principles of international environmental law include:

— the precautionary principle

— the polluter pays principle

— the principle of sustainable development (Brundtland Report, WSSD) - integration of environmental protection and economic development

— environmental procedural rights

— common but differentiated responsibilities

— intergenerational and intragenerational equity

— common concern of humankind

— common heritage

— partnership (WSSD)

— requirement to conduct a comprehensive environmental impact assessment

The Earth Summit

The Earth Summit in Rio de Janeiro was unprecedented for a UN conference, in terms of both its size and the scope of its concerns. Twenty years after the first global environment conference, the UN sought to help Governments rethink economic development and find ways to halt the destruction of irreplaceable natural resources and pollution of the planet. Hundreds of thousands of people from all walks of life were drawn into the Rio process. They persuaded their leaders to go to Rio and join other nations in making the difficult decisions needed to ensure a healthy planet for generations to come.

The Summit's message - that nothing less than a transformation of our attitudes and behaviour would bring about the necessary changes - was transmitted by almost 10,000 on-site journalists and heard by millions around the world. The message reflected the complexity of the problems facing us: that poverty as well as excessive consumption by affluent populations place damaging stress on the environment. Governments recognized the need to redirect international and national plans and policies to ensure that all economic decisions fully took into account any environmental impact. And the message has produced results, making eco-efficiency a guiding principle for business and governments alike.

— Patterns of production - particularly the production of toxic components, such as lead in gasoline, or poisonous waste - are being scrutinized in a systematic manner by the UN and Governments alike;
— Alternative sources of energy are being sought to replace the use of fossil fuels which are linked to global climate change;
— New reliance on public transportation systems is being emphasized in order to reduce vehicle emissions, congestion in cities and the health problems caused by polluted air and smog;
— There is much greater awareness of and concern over the growing scarcity of water.

The two-week Earth Summit was the climax of a process, begun in December 1989, of planning, education and negotiations among all Member States of the United Nations, leading to the adoption of Agenda 21, a wide-ranging blueprint for action to achieve sustainable development worldwide. At its close, Maurice Strong, the Conference Secretary-General, called the Summit a "historic moment for humanity". Although Agenda 21 had been weakened by compromise and negotiation, he said, it was still the most comprehensive and, if implemented, effective programme of action ever sanctioned by the international community. Today, efforts to ensure its proper implementation continue, and they will be reviewed by the UN General Assembly at a special session to be held in June 1997.

The Earth Summit influenced all subsequent UN conferences, which have examined the relationship between human rights, population, social development, women and human settlements - and the need for environmentally sustainable development. The World Conference on Human Rights, held in Vienna in 1993, for example, underscored the right of people to a healthy environment and the right to development, controversial demands that had met with resistance from some Member States until Rio.

The relationship between economic development and environmental degradation was first placed on the international agenda in 1972, at the UN Conference on the Human Environment, held in Stockholm. After the Conference, Governments set up the United Nations Environment Programme (UNEP), which today continues to act as a global catalyst for action to protect the environment. Little, however, was done in the succeeding years to integrate environmental concerns into national economic planning and decision-making. Overall, the environment continued to deteriorate, and such problems as ozone depletion, global warming and water pollution grew more serious, while the destruction of natural resources accelerated at an alarming rate.

By 1983, when the UN set up the World Commission on Environment and Development, environmental degradation, which had been seen as a side effect of industrial wealth with only a limited impact, was understood to be a matter of survival for developing nations. Led by Gro Harlem Brundtland of Norway, the Commission put forward the concept of sustainable development as an alternative approach to one simply based on economic growth - one "which meets the needs of the present without compromising the ability of future generations to meet their own needs".

After considering the 1987 Brundtland report, the UN General Assembly called for the UN Conference on Environment and Development (UNCED). The primary goals of the Summit were to come to an understanding of "development" that would support socio-economic development and prevent the continued deterioration of the environment, and to lay a foundation for a global partnership between the developing and the more industrialized countries, based on mutual needs and common interests, that would ensure a healthy future for the planet.

The Earth Summit Agreements

In Rio, Governments - 108 represented by heads of State or Government -

adopted three major agreements aimed at changing the traditional approach to development:

— Agenda 21 - a comprehensive programme of action for global action in all areas of sustainable development;

— The Rio Declaration on Environment and Development - a series of principles defining the rights and responsibilities of States;

— The Statement of Forest Principles - a set of principles to underlie the sustainable management of forests worldwide.

In addition, two legally binding Conventions aimed at preventing global climate change and the eradication of the diversity of biological species were opened for signature at the Summit, giving high profile to these efforts:

— The United Nations Framework Convention on Climate Change

and

— The Convention on Biological Diversity

Agenda 21 addresses today's pressing problems and aims to prepare the world for the challenges of the next century. It contains detailed proposals for action in social and economic areas (such as combating poverty, changing patterns of production and consumption and addressing demographic dynamics), and for conserving and managing the natural resources that are the basis for life - protecting the atmosphere, oceans and biodiversity; preventing deforestation; and promoting sustainable agriculture, for example.

Governments agreed that the integration of environment and development concerns will lead to the fulfilment of basic needs, improved standards for all, better protected and better managed ecosystems and a safer and a more prosperous future. "No nation can achieve this on its own. Together we can - in a global partnership for sustainable development", states the preamble.

The programme of action also recommends ways to strengthen the part played by major groups - women, trade unions, farmers, children and young people, indigenous peoples, the scientific community, local authorities, business, industry and non-

The Rio Declaration on Environment and Developmentsupports Agenda 21 by defining the rights and responsibilities of States regarding these issues. Among its principles:

— That human beings are at the centre of concerns for sustainable development. They are entitled to a healthy and productive life in harmony with nature;

— That scientific uncertainty should not delay measures to prevent environmental degradation where there are threats of serious or irreversible damage;

— That States have a sovereign right to exploit their own resources but not to cause damage to the environment of other States;

— That eradicating poverty and reducing disparities in worldwide standards of living are "indispensable" for sustainable development;

— That the full participation of women is essential for achieving sustainable development; and

— That the developed countries acknowledge the responsibility that they bear in the international pursuit of sustainable development in view of the pressures their societies place on the global environment and of the technologies and financial resources they command.

The Statement of Forest Principles, the non-legally binding statement of principles for thc sustainable management of forests, was the first global consensus reached on forests. Among its provisions:

— That all countries, notably developed countries, should make an effort to "green the world" through reforestation and forest conservation;

— That States have a right to develop forests according to their socio-economic needs, in keeping with national sustainable development policies; and

— That specific financial resources should be provided to develop programmes that encourage economic and social substitution policies.

At the Summit, the UN was also called on to negotiate an international legal agreement on desertification, to hold talks on preventing the depletion of certain fish stocks, to devise a programme of action for the sustainable development of small island developing States and to establish mechanisms for ensuring the implementation of the Rio accords.

The Earth Summit succeeded in presenting new perspectives on economic progress. It was lauded as the beginning of a new era and its success would be measured by the implementation - locally, nationally and internationally - of its agreements. Those attending the Summit understood that making the necessary changes would not be easy: it would be a multi-phased process; it would take place at different rates in different parts of the world; and it would require the expenditure of funds now in order to prevent much larger financial and environmental costs in the future.

In Rio, the UN was given a key role in the implementation of Agenda 21. Since then, the Organization has taken steps to integrate concepts of sustainable development into all relevant policies and programmes. Income-generating projects increasingly take into account environmental consequences. Development assistance programmes are increasingly directed towards women, given their central roles as producers and as caretakers of families. Efforts to manage forests in a sustainable manner begin with finding alternatives to meet the needs of people who are overusing them. The moral and social imperatives for alleviating poverty are given additional urgency by the recognition that poor people can cause damage to the environment. And foreign investment decisions increasingly take into account the fact that drawing down the earth's natural resources for short-term profit is bad for business in the long run.

In adopting Agenda 21, the Earth Summit also requested the United Nations to initiate talks aimed at halting the rapid depletion of certain fish stocks and preventing conflict over fishing on the high seas. After negotiations spanning more than two years, the UN Agreement on High Seas Fishing was opened for signature on 4 December 1995. It provides for all species of straddling and highly migratory fish - those which swim between national economic zones or migrate across broad areas of the ocean - to be subject to quotas designed to ensure the continued survival of fish for our children and grandchildren to enjoy.

Also at the Summit, Governments requested the UN to hold negotiations for an international legal agreement to prevent the degradation of drylands. The resulting International Convention to Combat Desertification in Those Countries Experiencing Serious Drought and/or Desertification, particularly in Africa, was opened for signing in October 1994 and entered into force in December 1996. It calls for urgent action to be taken in Africa, where some 66 per cent of the continent is desert or drylands and 73 per cent of agricultural drylands are already degraded.

In order to promote the well-being of people living in island countries, the Summit called for the UN to convene a Global Conference on the Sustainable Development of Small Island Developing States . The

Conference was held in Barbados in May 1994 and produced a programme of action designed to assist these environmentally and economically vulnerable countries.

In addition, three bodies were created within the United Nations to ensure full support for implementation of Agenda 21 worldwide:

— The UN Commission on Sustainable Development, which first met in June 1993;

— The Inter-agency Committee on Sustainable Development, set up by the Secretary-General in 1992 to ensure effective system-wide cooperation and coordination in the follow-up to the Summit; and

— The High-level Advisory Board on Sustainable Development, established in 1993 to advise the Secretary-General and the Commission on issues relating to the implementation of Agenda 21.

UN Commission on Sustainable Development (CSD) - The Earth Summit called on the General Assembly to establish the Commission under the Economic and Social Council as a means of supporting and encouraging action by Governments, business, industry and other non-governmental groups to bring about thc social and economic changes needed for sustainable development. Each year, the Commission reviews implementation of the Earth Summit agreements, provides policy guidance to Governments and major groups involved in sustainable development and strengthens Agenda 21 by devising additional strategies where necessary. It also promotes dialogue and builds partnerships between Governments and the major groups which are seen as key to achieving sustainable development worldwide. The work of the Commission was supported by numerous inter-sessional meetings and activities initiated by Governments, international organizations and major groups. In June 1997, the General Assembly will hold a special session to review overall progress following the Earth Summit.

Under a multi-year thematic work programme, the Commission has monitored the early implementation of Agenda 21 in stages. Each sectoral issue - health, human settlements, freshwater, toxic chemicals and hazardous waste, land, agriculture, desertification, mountains, forests, biodiversity, atmosphere, oceans and seas - was reviewed between 1994 and 1996. Developments on most "cross-sectoral" issues are considered each year. These issues, which must be addressed if action in sectoral areas is to be effective, are clustered as follows: critical elements of sustainability (trade and environment, patterns of production and consumption, combating poverty, demographic dynamics); financial resources and mechanisms; education, science, transfer of environmentally sound technologies, technical cooperation and capacity-building; decision-making; and activities of the major groups, such as business and labour. (For further details click here on UNCSD.)

In 1995, the Commission established under its auspices the Intergovernmental Panel on Forests with a broad mandate covering the entire spectrum of forest-related issues and dealing with conservation, sustainable development and management of all types of forests. The Panel will submit its final report containing concrete conclusions and proposals for action to the 1997 session of the CSD. (For further details click here on IPF)

Reports submitted annually by Governments are the main basis for monitoring progress and identifying problems faced by countries. By mid-1996, some 100 Governments had established national sustainable development councils or other coordinating bodies. More than 2,000 municipal and town governments had each formulated a local Agenda 21 of its own. Many countries were seeking legislative approval for sustainable development plans, and the level of NGO involvement remained high.

Convention on International Trade in Endangered Species of wild fauna and flora (CITES), 1973

The aim of CITES is to control or prevent international commercial trade in endangered species or products derived from them. CITES does not seek to directly protect endangered species or curtail development practices that destroy their habitats. Rather, it seeks to reduce the economic incentive to poach endangered species and destroy their habitat by closing off the international market. India became a party to the CITES in 1976. International trade in all wild flora and fauna in general and species covered under CITES is regulated jointly through the provisions of The Wildlife (Protection) Act 1972, the Import/Export policy of Government of India and the Customs Act 1962 (Bajaj, 1996).

Montreal Protocol on Substances that deplete the Ozone Layer (to the Vienna Convention for the Protection of the Ozone Layer), 1987

The Montreal Protocol to the Vienna Convention on Substances that deplete the Ozone Layer, came into force in 1989. The protocol set targets for reducing the consumption and production of a range of ozone depleting substances (ODS). In a major innovation the Protocol recognized that all nations should not be treated equally. The agreement acknowledges that certain countries have contributed to ozone depletion more than others. It also recognizes that a nation's obligation to reduce current emissions should reflect its technological and financial ability to do so. Because of this, the agreement sets more stringent standards and accelerated phase-out timetables to countries that have contributed most to ozone depletion (Divan and Rosencranz, 2001).

India acceded to the Montreal Protocol along with its London Amendment in September 1992. The MoEF has established an Ozone Cell and a steering committee on the Montreal Protocol to facilitate implementation of the India Country Program, for phasing out ODS production by 2010.

To meet India's commitments under the Montreal Protocol, the Government of India has also taken certain policy decisions.

— Goods required to implement ODS phase-out projects funded by the Multilateral Fund are fully exempt from duties. This benefit has been also extended to new investments with non-ODS technologies.
— Commercial banks are prohibited from financing or refinancing investments with ODS technologies.

The Gazette of India on 19 July 2000 notified rules for regulation of ODS phase-out called the Ozone Depleting Substances (Regulation and Control) Rules, 2000. They were notified under the Environment (Protection) Act, 1986. These rules were drafted by the MoEF following consultations with industries and related government departments.

Basel Convention on Transboundary Movement of Hazardous Wastes, 1989

Basel Convention, which entered into force in 1992, has three key objectives:

— To reduce transboundary movements of hazardous wastes;
— To minimize the creation of such wastes; and
— To prohibit their shipment to countries lacking the capacity to dispose hazardous wastes in an environmentally sound manner.

India ratified the Basel Convention in 1992, shortly after it came into force. The Indian Hazardous Wastes Management Rules Act 1989, encompasses some of the Basel provisions related to the notification of import and export of hazardous waste, illegal traffic, and liability.

UN Framework Convention on Climate Change (UNFCCC), 1992

The primary goals of the UNFCCC were to stabilize greenhouse gas emissions at levels that would prevent dangerous anthropogenic interference with the global climate. The convention embraced the principle of common but differentiated responsibilities which has guided the adoption of a regulatory structure.

India signed the agreement in June 1992, which was ratified in November 1993. As per the convention the reduction/limitation requirements apply only to developed countries. The only reporting obligation for developing countries relates to the construction of a GHG inventory. India has initiated the preparation of its First National Communication (base year 1994) that includes an inventory of GHG sources and sinks, potential vulnerability to climate change, adaptation measures and other steps being taken in the country to address climate change. The further details on UNFCC and the Kyoto Protocol are provided in Atmosphere and climate chapter.

Convention on Biological Diversity, 1992

The Convention on Biological diversity (CBD) is a legally binding, framework treaty that has been ratified until now by 180 countries. The CBD has three main thrust areas: conservation of biodiversity, sustainable use of biological resources and equitable sharing of benefits arising from their sustainable use.

The Convention on Biological Diversity came into force in 1993. Many biodiversity issues are addressed in the convention, including habitat preservation, intellectual property rights, biosafety, and indigenous peoples' rights.

UN Convention on Desertification, 1994

Delegates to the 1992 UN Conference on Environment and Development (UNCED) recommended establishment of an intergovernmental negotiating committee for the elaboration of an international convention to combat desertification in countries experiencing serious drought and/or desertification. The UN General Assembly established such a committee in

1992 that later helped formulation of Convention on Desertification in 1994.The convention is distinctive as it endorses and employs a bottom-up approach to international environmental cooperation. Under the terms of the convention, activities related to the control and alleviation of desertification and its effects are to be closely linked to the needs and participation of local land-users and non-governmental organizations. Seven countries in the South Asian region are signatories to the Convention, which aims at tackling desertification through national, regional and sub-regional action programmes. The Regional Action Programme has six Thematic Programme Networks (TPN's) for the Asian region, each headed by a country task manager. India hosts the network on agroforestry and soil conservation. For details refer to the land resource chapter.

International Tropical Timber Agreement and The International Tropical Timber Organisation (ITTO), 1983, 1994

The ITTO established by the International Tropical Timber Agreement (ITTA), 1983, came into force in 1985 and became operational in 1987. The ITTO facilitates discussion, consultation and international cooperation on issues relating to the international trade and utilization of tropical timber and the sustainable management of its resource base. The successor agreement to the ITTA (1983) was negotiated in 1994, and came into force on 1 January 1997. The organization has 57 member countries. India ratified the ITTA in 1996.

References

Davic, R. D. (2003). "Linking keystone species and functional groups: a new operational definition of the keystone species concept". *Conservation Ecology* 7 (1): r11.

Mills, L. S.; Soule, M. E.; Doak, D. F. (1993). "The keystone-species concept in ecology and conservation".*BioScience* 43 (4): 219–224.

McCann, K. (2007). "Protecting biostructure". *Nature* 446 (7131): 29. Bibcode 2007Natur.446...29M.

Noss, R. F.; Carpenter, A. Y. (1994). *Saving Nature's Legacy: Protecting and Restoring Biodiversity*. Island Press. p. 443.

8

Environmental Education and Awareness

Environmental Education (EE) refers to organized efforts to teach about how natural environments function and, particularly, how human beings can manage their behavior and ecosystems in order to live sustainably. The term is often used to imply education within the school system, from primary to post-secondary. However, it is sometimes used more broadly to include all efforts to educate the public and other audiences, including print materials, websites, media campaigns, etc. Related disciplines include outdoor education and experiental education.

Environmental education is a learning process that increases people's knowledge and awareness about the environment and associated challenges, develops the necessary skills and expertise to address the challenges, and fosters attitudes, motivations, and commitments to make informed decisions and take responsible action.

The roots of environmental education can be traced back as early as the 18th century when Jean-Jacques Rousseau stressed the importance of an education that focuses on the environment in Emile: or, On Education. Several decades later, Louis Agassiz, a Swiss-born naturalist, echoed Rousseau's philosophy as he encouraged students to "Study nature, not books." These two influential scholars helped lay the foundation for a concrete environmental education program, known as Nature study, which took place in the late 19th century and early 20th century.

The nature study movement used fables and moral lessons to help students develop an appreciation of nature and embrace the natural world.

Anna Botsford Comstock, the head of the Department of Nature Study at Cornell University, was a prominent figure in the nature study movement and wrote the Handbook for Nature Study in 1911, which used nature to educate children on cultural values. Comstock and the other leaders of the movement, such as Liberty Hyde Bailey, helped Nature Study garner tremendous amounts of support from community leaders, teachers, and scientists and change the science curriculum for children across the United States.

A new type of environmental education, Conservation Education, emerged as a result of the Great Depression and Dust Bowl during the 1920s and 1930s. Conservation Education dealt with the natural world in a drastically different way from Nature Study because it focused on rigorous scientific training rather than natural history. Conservation Education was a major scientific management and planning tool that helped solve social, economic, and environmental problems during this time period.

The modern environmental education movement, which gained significant momentum in the late 1960s and early 1970s, stems from Nature Study and Conservation Education. During this time period, many events – such as Civil Rights, the Vietnam War, and the Cold War – placed Americans at odds with one another and the U.S. government. However, as more people began to fear the fallout from radiation, the chemical pesticides mentioned in Rachel Carson's Silent Spring, and the significant amounts of air pollution and waste, the public's concern for their health and the health of their natural environment led to a unifying phenomenon known as environmentalism.

The first article about environmental education as a new movement appeared in Phi Delta Kappan in 1969, authored by James A. Swan. A definition of "Environmental Education" first appeared in Educational Digest in March 1970, authored by William Stapp Stapp later went on to become the first Director of Environmental Education for UNESCO, and then the Global Rivers International Network.

Ultimately, the first Earth Day on April 22, 1970 – a national teach-in about environmental problems – paved the way for the modern environmental education movement. Later that same year, President Nixon passed the National Environmental Education Act, which was intended to incorporate environmental education into K-12 schools. Then, in 1971, the National Association for Environmental Education (now known as the North

American Association for Environmental Education) was created to improve environmental literacy by providing resources to teachers and promoting environmental education programs.

Internationally, environmental education gained recognition when the UN Conference on the Human Environment held in Stockholm, Sweden, in 1972, declared environmental education must be used as a tool to address global environmental problems. The United Nations Education Scientific and Cultural Organization (UNESCO) and United Nations Environment Program (UNEP) created three major declarations that have guided the course of environmental education.

The Concept Environmental Education

In order to define the concept "environmental education" one needs to focus on the meaning of "education" and "environment". The term "education" can be defined as the field of study that is concerned with the pedagogy of teaching and learning. The learning process should produce an output while the total development of the individual is the focus of education. Education is a gradual, intangible process and includes the inculcation of values.

The environment is everything around us. It includes people themselves, their history, where they live, learn, work, play, relax and enjoy nature. Environmental resources are used to support life systems. These resources can be natural such as the soil, water and air, or they can be man-made such as houses, bridges and vehicles. These natural resources form the life support system and should thus be used wisely.

Environmental Education (EE) has always been seen as the solution to environmental degradation. The definition and conceptualization of how EE should occur has, however, changed over the last number of decades. Initially it was actually viewed as ecological education with little consideration given to human needs. It was also located within certain disciplines, which were thought to be the appropriate vehicles for EE.

Today EE is viewed more holistically and is considered to be an interdisciplinary consideration. EE is regarded as the acquisition of a critical understanding of human-ecological principles and pr ocesses (i.e. socio-economic and biophysical) to make informed decisions about environmental issues. It enables the exploration of environmental issues through educational experiences and reflection in the environment. Education would be progressive if people see the environment as something to be cared for themselves.

Definitions of EE vary. Some of the definitions of EE found on the internet are:

— "Activities with organized groups or seminar participants that are designed to develop understanding, appreciation, and caring for the natural environment".

— "Positions may be found at arboreta or botanical gardens, nature centers, wildlife centers, camps, parks, alternative schools, and working with special populations. Interns learn teaching techniques, teach about nature, lead nature walks, lead outdoor pursuits etc.

— "The study and analysis of the conditions and causes of pollution, overpopulation, and waste of natural resources, and of the ways to preserve Earth's intricate ecology." highered.

EE generally refers to curriculum and programmes that aim to teach people about the natural world and particularly about ways in which ecosystems work. EE programmes often aim to change people's perceptions about the value of the natural world and to teach them how to change environmental behaviours, such as getting people to recycle or how to build eco- friendly dwellings. In Western-style EE the focus is mostly on understanding ways in which humans and human systems impact on the environment and non-human natural systems. However, it is also important to teach and understand the impact of natural systems on humans and human society.

Janse van Rensburg and Lotz refer to EE as " ... a process through which we might enable ourselves and future generations to respond to environmental issues in ways that might foster change towards sustainable community life in a healthy environment". According to Blignaut EE is a " ... process which enables people to understand the part they play in reducing the tensions between dichotomies in order to achieve the best possible balance". The dichotomies that are referred to here are: economics and technology; education for individual development and human rights; ecology; health and conservation education; peace education for constructive community interaction; and the social aspect.

Many scholars view EE as embracing "'society' and the 'environment' in an inseparable process". It is also described as involving an "... understanding of political processes and the creation of political and legal structures, which enable the individual to participate actively in decision making about environmental issues on a local, national and global scale". Furthermore, EE is viewed as a " ... process or approach, which develops correct attitudes, values, behaviour and skills which will enable people to live in harmony with the natural resources, and maintain a good quality of life considering that there are still generations to come".

The EE Association of South Africa (EEASA) promotes the idea that "... we share one environment, and the better we share it and collectively care for it, the better future all of us are likely to have". EEASA proposes that adoption of the principle that "...EE is a fundamental necessity for a successful society, espousing principles or values such as democratization, respect for people and the natural resources". According to Meadows the UNESCO -UNEP Education Series stipulates that some of the key principles of EE are :

— levels of being, i.e. the physical planet, the biosphere, the sociosphere and the technosphere;
— cycles;
— complex systems;
— population growth and carrying capacity;
— sustainable development; and
— knowledge and uncertainty.

EE thus implies a balanced growth in appreciation and understanding of the environment and responsible behaviour in all its dimensions. Baez, et al support this view when they include the spatial, social and temporal environment, as well as one's own internal environment, which has an influence on the behaviour of both physical and physiological needs, and of memory, imagination and creativity.

Tselane and Mosidi describe EE as a process that develops correct attitudes, values, behaviour and skills. These will enable pe ople to live harmoniously with their natural resources and maintain a good quality of life. It will also support the exploration of environmental issues through educational experiences and a reflection on the environment, knowledge about the environment, and appropriate actions with regard to the environment. They further state that to maintain harmony between the natural resources and people, each person's attitudes and actions need careful consideration. It is therefore envisaged that environmental considerations and engagement in the new education system should not be negotiable.

Based on the above descriptions, EE can be viewed as an attempt to cause learners to appreciate, understand and have a basic knowledge of the environment. It entails the development of values, attitudes and skills regarding responsible behaviour on the part of the present generation, bearing future generations in mind. It also aims at the judicious utilization of existing resources so that future generations can enjoy a comfortable way of life as is currently experienced in most parts of the world. The following points characterize EE:

— EE includes a human component in the exploration of environmental problems and solutions.
— EE rests on a foundation of knowledge about social and ecological systems.
— EE includes the affective domain: attitudes, values and commitments necessary to build a sustainable society.
— EE includes opportunities to build skills that enhance learners' problem -solving abilitics.

Many references to EE acknowledge that the Bill of Rights in the New Constitution of South Africa enshrines "the right of every citizen to an environment, which is not detrimental to his or her health". This was also proposed in the discussion document of April 1998 on "Enabling EE as a cross-curricular concern in Outcomes -Based Learning Programmes", which states that the prevention of environmental degradation, solutions to problems and development of sustainable living need to be addressed. Vital programmes of learning significant for South Africa's development should include environmental concerns, viewed from a socio-ecological and soc io-historical perspective, which emphasize sustainable development and management of the life- sustaining support systems.

Curriculum Models

Many researchers prefer not to refer to EE as a "discipline" as its holistic nature is then not accounted for. EE is rather considered to be "interdisciplinary" rather than "multidisciplinary" because of its complex nature and reliance on other disciplines, such as Science, Mathematics, Geography and Home Economics. Two models have been utilized in the curriculum development and implementation of EE. The first model focuses on EE as the creation of a single subject where EE is referred to as an interdisciplinary approach, whilst, secondly, the incorporation of an EE component into other disciplines is referred to as a multidisciplinary (infusion) approach. These two models are illustrated in figure 1.

The interdisciplinary approach emphasizes curriculum organization and/or teaching approaches which cut across subject boundaries to focus on general concepts and life-skills, as well as comprehensive environmental problems and issues. It also focuses on broad study areas that bring together the various segments of the curriculum into meaningful association. The multidisciplinary approach refers to a "curriculum organization" where contents from one subject discipline are aligned with the concepts/contents of another, at the same time ma intaining clear boundaries between disciplines. The integrative effect lies on a continuum from more to less integration, depending on the focus of the multidisciplinary exercise.

Development of EE

The Belgrade Charter was one of the initial documents to give considerable attention to EE. Leal Filho states that t h e goal of EE is "to develop a world population that is aware of, and concerned about the environment, and its associated problems, and which has the knowledge, skills, attitudes, motivations and commitment to work individually and collectively toward the solutions of current problems and the prevention of new ones".

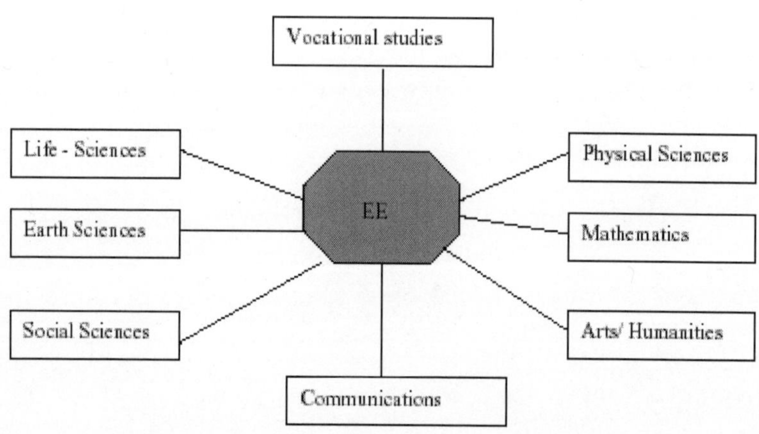

a: Interdisciplinary Model (single subject approach)

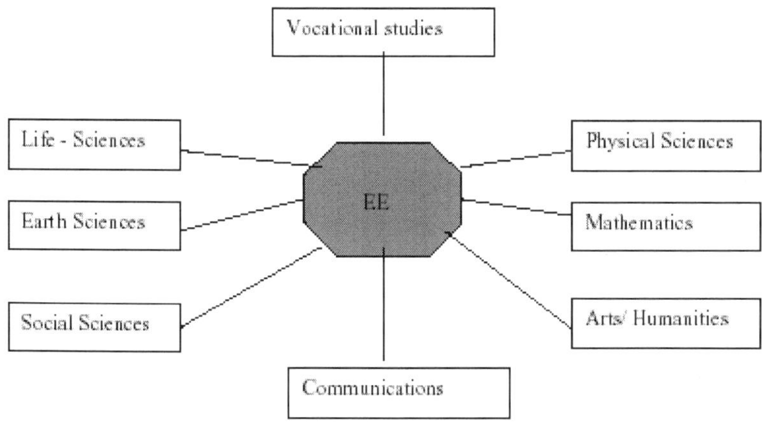

b: Multidisciplinary Model (infusion approach)

Figure 1. Curriculum models for environmental education

Environment has mostly been included in school programmes in one form or another. This is because the environment provides a rich and convenient source of examples for studying various subjects. EE therefore dates back to Aristotle and Rousseau, while educational consideration of the natural world has mostly included attention to aesthetic qualities and utilitarian aspects.

Hungerford and Peyton state that as an independent entity EE is of fairly recent origin. However, its roots extend to when man first envisioned an interrelationship between him/herself and the biosphere. This relationship resulted in an evaluation of his/her role in maintaining or damaging the environment. EE only emerged as a field of concern during the mid- 1960's. However, in the 1930's progressive education fostered "learning by doing", thus emphasizing an interdisciplinary approach to education as opposed to a multidisciplinary approach. This incorporated learning about the environment in the environment. It fostered holistic, integrated, interdisciplinary education. In the 1960's, concerns were voiced for the environmental quality of human life. Since the late 1960's, EE has been regarded as education focused on the environment. The primary antecedents were nature study, outdoor study and conservation education.

The development of explicit interconnections between human health, science and technology and the environmental, economic and social issues and problems of society has characterized EE since the 1970's. In 1972, Disinger initiated the idea of "in/about/for" the environment.

More recently, the roots of EE have been traced back to the conservation movement. Its roots and strengths lie in conservation education, nature education, resource- use education, outdoor education, geography education and science education. Differences do, however, exist between conservationism and environmentalism.

The focus of EE can be derived from:

— the interrelationships between natural and social systems;

— the unity of humankind with nature;

— technology and the making of choices; and

— development learning throughout the hum an life cycle.

Disinger states that EE focuses on five clusters of values. The five clusters are natural heritage; public health and the environment; careers; sustainable development and quality education. These clusters bring into focus the necessary relationship of the environment to both the national and the educational mainstreams - they reflect the values espoused by the society.

Recently, and because of agreement that our environmental crises are the res ult of problems with modernity, teachers have begun to move away from teaching approaches that stress 'wildlife experience' or 'nature study'. Teachers now prefer an approach that encourages learners to understand and transform problem environments. It is in this sense that we now prefer to speak of EE as education in, about and for the environment. Ten years ago EE would have been equated with the environmental sciences, a field which is dominated by the conventions and traditions of the scientific method. EE is now rather seen as a holistic field that draws from the tools of both the social and natural sciences.

Education for and about the Environment

Nowadays, the focus has moved away from only education about (knowledge) and in and/or through the environment (direct encounter and experiences in nature) to include education for t h e environment (aimed strongly at attitudes and action competencies). Fien and Van Rooyen assert that the rationale and methodology of environmental teaching assumes an interdisciplinary approach. It focuses on education about, through and for the environment, and can be described as:

— Education about the environment is the most common form of EE. Its objectives emphasize knowledge about the natural systems and processes; development of environmental investigation and thinking skills, and understanding the ecological, economic and political, scientific- technological and socio-cultural factors that influence peoples' decision- making. Concern needs to be translated into appropriate behaviour patterns and actions, but for this to happen, it is essential for learners to understand how the natural systems work, as well as the impact of human activities upon them. Many non-formal and formal avenues of EE, including the arts, the natural and social sciences can provide such knowledge.

— Education in or through the environment makes use of learners' own experiences. It aims to add reality, relevance and practical experience to learning, and to provide learners with an aesthetic appreciation of the environment through contact.

— Education for the environment focuses on the values of education and social change. Its aim is to engage learners in the exploration and resolution of environmental issues in order to promote lifestyles that are compatible with the sustainable and equitable use of resources. This approach builds on education about and through the environment, the development of an informed concern for the environment, a sensitive environmental ethic, the motivation and skills for participating in protecting and improving the environment, and promoting a willingness and ability to adopt lifestyles compatible with the wise use of environmental resources. EE is basically composed of aspects including about, in and/or through and for the environment.

Education for Sustainability

Since the Earth Summit, sustainable development has been a key theme at a series of United Nations conference s discussing pathways to development.These conferences have shown that the interdependent links between environment and development are not simply about conservation and economics, but also include a concern for issues such as human rights, population, housing, food security, and gender that are important parts of sustainable human development. The following conference themes are an indication of these concerns:

— Human Rights - World Conference on Human Rights - Vienna, Austria, 1993

— Population - International Conference on Population and Development - Cairo, Egypt, 1994

— Social Development - World Summit for Social Development - Copenhagen, Denmark, 1995

— Women - Fourth World Conference on Women - Beijing, People's Republic of China, 1995

— Housing and Settlements - Second United Nations Conference on Human Settlements (Habitat II) - Istanbul, Turkey, 1996

— Food Security - World Food Summit - Rome, Italy 1996

Action on these quality of life issues has been reviewed in a series of follow-up conferences:

— Environment and Development - Rio, 1997 and World Conference on Sustainable Development - Johannesburg, 2002

— Human Rights - Vienna, 1998

— Population - Cairo, 1999

— Social Development - Copenhagen, 2000

— Women - Beijing, 2000

— Housing and Settlements - Istanbul, 2001

— Food Security- Rome, 2001

These conferences provided an opportunity for the international community to start discussing a Comprehensive Development Framework and, eventually, to agree on a set of international development goals. These initiatives take a holistic approach to development in which there is a balance between all dimensions of development - social, economic, political and ecological. Recently, there have been a few attempts at defining the characteristics of this latest approach to education. The report of the British Environment and Development Education and Training Group (EDET), Good Earth-Keeping: Education, Training and Awareness for a Sustainable Future, defines the nature of education for sustainability as follows:

> "We believe that education for sustainability is a process which is relevant to all people and that, like sustainable development itself, it is a proc ess rather than a fixed goal. It may precede - and it will always accompany -

the building of relationships between individuals, groups and their environment. All people, we believe, are capable of being teachers and learners in pursuit of sustainability".

In its report, the British EDET Group affirmed the validity of the different approaches to environmental education in achieving sustainable development. However, Tilbury and Fien argue that environmental education for sustainability must differ significantly from the apolitical, naturalist and scientific work carried out under the environmental education banner of the 1970's and 1980's. Education with the objective of achieving sustainability varies from previous approaches to EE in that it focuses more sharply on developing closer links between environmental quality, ecology and socio-economics and the political threads that underlie these.

To this end, several emphases in education that result from this integra tive view of education and sustainability are now being explored. These relate to the special need (1) to consciously link studies of the natural and social worlds, (2) to emphasise environmental citizenship, (3) to develop ethics and values, and (4) to incorporate a futures perspective in our programmes. Each of these emphases of education for sustainability is accompanied by several questions for reflection on possible implications for education in botanic gardens.

Importance of Environmental Literacy

Environmental literacy is a common goal of general education. Since 1989 an EE task force of the American Society for Testing and Mater ials has sought the establishment of consensus for EE Participation. This task force involves experienced, active and environmentally literate individuals in EE. Over time environmental literacy came to be viewed as the primary goal of EE. Various perceptions of environmental literacy have evolved as more people became involved in this field. In keeping with EE, we can say that environmental literacy:

— is the essential capacity to perceive and interpret the relative health of environmental systems, and citizens therefore take appropriate action to maintain, restore, or improve the health of such systems;
— develops with the objective of fostering productive and responsible citizens of this planet and its societies;
— develops regular and continuous involvement throughout schooling and should also form part of the basic core programme of schools; and

— develops and fosters a key objective of any general education programme.

There has been much research in determining how EE affects environmental literacy. Many teachers view environmental literacy in terms of the knowledge, skills and behaviours a student should be able to demonstrate when schooling is completed. According to the Rio Declaration, literacy guidelines are based on the assumption that an environmentally literate person should possess:

— an awareness and sensitivity to the total environment;
— a variety of experiences in and a basic understanding of environmental problems;
— a set of environmental values and a feeling of concern for the environment, as well as the motivation and disposition to actively participate in environmental improvement and protection; and
— skills for identifying, investigating and solving environmental problems.

Environmental problems could be controlled if a nation can control population growth, judiciously manage resources, improve the quality of the atmosphere and prevent further extinction of plant and animal species. We must begin to educate the next generation for a "quality environment" in order to balance their needs of nature and human populations whenever the need for decision- making arises.

Levels of Environmental Literacy

Environmental literacy can be defined as the ability to demonstrate observably what has been learned, i.e. knowledge of key concepts, acquired skills and disposition towards issues. It builds on an ecological paradigm and is the capacity to perceive and interpret the relative health of environmental systems and to take appropriate action to maintain, restore or improve the health of those systems. Environmental literacy is a continuum of competencies ranging from zero to high competency. These can further be divided into three working levels:

— The nominal level indicates the ability to recognize many of the basic terms used in communicating about the envir onment and to provide rough, if unsophisticated, working definitions of their meanings.
— The functional level indicates a broader knowledge and understanding of the nature and interactions between human and social systems and other natural systems.

— The operational level is the progress beyond functional literacy in both the breadth and depth of understandings and skills.

The development of levels has been further divided into strands in environmental literacy.

Strands in Environmental Literacy

Within the above three major levels there are four major strands. According to Roth these are:

— the knowledge strand,

— the skills strand,

— the affect strand, and

— the behaviour strand.

The knowledge strand refers to the familiarity and understanding of systems and nature. The affect strand relates to feelings, values, attitudes and an appreciation of nature and society. The skills strand refers to skills of identification, analysis, synthesis, application of suitable action, evaluation of actions implemented, risk taking, decision making and critical thinking. The behavioural strand refers to the actions implemented, either individually or collectively. This idea is reiterated by the North American Association of Environmental Educ ation (NAAEE). It too believes that environmental literacy means "understanding how human actions and decisions affect environmental quality and acting on that understanding in a responsible and effective manner. Informed and active participation is a key component of environmental literacy and a common goal of service learning and citizenship education. In fact, all four of the major aspects of environmental literacy are applicable to, and necessary for, successful citizenship. This is especially true toda y when so many of our public and private decisions can affect public health and environmental quality.

Environmental literacy draws upon scientific literacy. It therefore involves people in:

— using critical and creative thinking;

— seeking and organizing information;

— being healthily skeptical; and

— thinking ahead and planning.

It has often been stated that environmental literacy draws upon six major areas namely:

— environmental sensitivity;
— knowledge;
— skills;
— attitudes and values;
— personal investment; and
— responsibility and active involvement.

The inculcation and development of strands and levels of literacy leads to the development of environmentally literate and active people.

Environmentally Literate and Active Citizenship

The creation of an environmentally literate citizen is the primary goal of EE. Although the term environmental literacy has been used since the 1960's and creates positive images of the environment - but it conveys little in terms of information and direction. The goal of EE is the development of an environmentally literate citizenry, i.e. a citizenry that is both competent to take action on critical environmental issues and willing to take action".

The attribute of an environmentally literate person is that he/she is at his/her highest level of competency, the operational level. Such people regularly evaluate the impact and consequences of actions, gather and synthesize pertinent information, choose among alternatives, advocate action positions, and take action after decision- making. The aim is to enhance a healthy environment. At this level people demonstrate a strong, ongoing sense of investment in and responsibility for preventing or remediating environmental degradation personally and collectively.

Action can be taken from a global to a local scope. In assessing the level of environmental literacy one needs to assess the efficiency of programmes to develop and nurture the acquired literacy. In schools it is therefore essential to test what is being demonstrated or not, and to take action to remediate deficiencies. It is advisable to use a checklist in such cases. Assessment requires the use of diverse formative assessment methods in order to develop the understanding, values, attitudes, skills and self - confidence required of an environmentally literate person".

Loubser used ten key concepts which he thought should be known/ understood by all environmentally literate individuals. These concepts correspond with concepts selected by other authors and institutions. The ten concepts are:

1. The earth as a closed system: the concept of ecosystem.
2. Human interaction within the environment: every action has an impact.
3. Cycles : natural
4. Interaction of economics, science and politics in EE: integration of learning about our world
5. Management of environme nt and resources for long-term sustainability.
6. Habitat: importance of food, water, shelter and space for personal/ human/animal survival.
7. Food webs and chains: biological magnification and contamination.
8. Complexity of decision-making on environmental issues.
9. Hope: Natural rehabilitation and regeneration from environmental damage.
10. Personal commitment for the care and respect of the environment.

Loubser states that in most of these concepts there is a close correlation between knowledge, skills, affect and behaviour. Loubser identified these concepts related to environmental literacy which he feels that teachers need to have a grasp on before they teach EE. Since EE is an important means to developing environmental literacy.

1. Basic understanding of the biosphere.
2. Understanding of an ecological perspective of nature and human beings.
3. Awareness of human interactions with the environment and interrelationships in an ecosystem.
4. Knowledge of environmental changes brought about by industrialization, urbanization.
5. Understanding of the activities to meet basic needs and wants and how it affects health, the environment and quality of life.
6. Awareness of renewable and non-renewable resources.
7. Knowledge of how to maintain environmental quality and qualit y of life.

8. An understanding about the ability to make choices.
9. Knowledge of decision making and environmental issues in scientific, economic, legal, social and political contexts.
10. Knowledge of environmental ethics as a way of life.

If the primary goal of EE is environmental literacy one may then ask what the underlying principles of EE are, especially in a formal curriculum. To help the reader understand this, the researcher decided to expand on the principles of EE.

Principles of Environmental Education

In Australia, the Department of Education identified a number of EE principles. These principles reinforce the position of EE in formal education. The Department stipulates tha t EE should be incorporated into formal education policy and into the curriculum at all levels. Environmental education is viewed as an educational process that should merge into formal education systems. It should, however, be regarded as a separate entity. Another principle is based on the holistic philosophy. Certain aspects should be studied within the context of interactions. The total environment should be considered from a balanced, integrated perspective. Curriculum structures should focus on an integrated and interdisciplinary approach.

The Department of Education in Queensland, Australia, emphasizes the use of diverse learning environments and a broad array of learner -centred approaches. The development of "sensitivity, empa thy, attitudes and values requires that the development of the affective domain be emphasized". The understanding of general and subject specific environmental concepts and the development of thinking, communication and action skills should be promoted.

In order to counteract negativism and promote positive action, skills of analytical and critical thinking should be integrated with synthesis and creative thinking in environmental problem -solving exercises. EE projects should be based on local community issues. Such projects develop and integrate communication and thinking skills acquired in the classroom. Opportunities should be provided for learners to become empowered by assisting in the planning of some of their own learning experiences, in making decisions and accepting the consequences of their actions. EE examines issues of local and global significance. It also makes use of a variety of teaching and learning strategies and resources. EE is therefore learner-centred, coherent and progressive, and includes a component of direct experience. It is inclusive, exemplary and community orientated.

Fien's principles of Environmental Education

Fien provided some ideas for schools and teachers to translate remote, centrally developed policies into meaningful goals and practices. At the subversive level he provided ideas for establishing socially critical practices in EE at the local level. In its social-critical form education can be viewed as "education for the environment". Such education seeks to promote a willingness and ability among learners to adopt lifestyles that are compatible with the socially and ecologically sustainable use of resources. This helps to build education in and about the environment. Acc ording to Fien the final report on the Tbilisi Conference stressed that EE be based upon a search for answers to a number of critical questions. Fien indicates that these answers will provide a proposed framework for curriculum enquiry, and that such a framework is neither exhaustive nor curriculum prescriptive. Countries throughout the world have approached EE differently. There have been two broad global approaches to the subject.

Approaches to Environmental Education

To a large extent EE goals around the world are similar, namely to maintain and improve environmental quality and to prevent future environmental problems. EE partly involves information education and increasing knowledge about the environment. It is also viewed as an increasing awareness about issues and an understanding of personal values by focusing on attitudes and values and by evaluating one's feelings regarding environmental issues.

Broadly speaking there are two approaches to EE, namely the North-American-European approach and the Australian approach. The latter is a socially-critical approach to EE, i.e. social issues feature prominently in EE. The trend of criticism lies mainly in social structures and ecosystems. Australians are critically minded when discussing the content of EE. The North-American-European approach has more positions on the continuum. This school of thought focuses on the holistic view of EE. The Australians are somewhat focused on and critical of social issues in their approach, whereas the North-American-European approach is more balanced and holistic.

Some EE authors consider the Australian policies as exemplary of current trends. It is often proposed that they have replaced earlier behaviourist and social engineering approaches. The Australians emphasize that the prevention and solving of environmental problems need methodology based on crit ical thinking. This implies that teaching will be characterized by co- operation, critical enquiry and interdisciplinary lessons. The classroom will represent a learner-environment focus, using social and cognitive theories. Curriculum development is partic ipatory, enquiry- based and critical in order to encourage dialogue, which leads to reconceptualisation and renewal. Clacherty quotes Greenall Gough when stating that Australian literature has moved towards socially critical EE. The cur riculum is issues - based and integrated with the rest of the general school curriculum. This approach to EE is critical and openminded and involves problem solving. Clacherty also states that the state of Victoria in Australia embarked on the most radical reformation of upper-secondary education regarding EE.

South African Approach

In South Africa, Curriculum 2005 reveals the approach to EE. This is an outcomes- based curriculum, underlined by a shift from the behaviourist approach. The emphasis is on the achievement of specific outcomes. In the behaviourist approach the emphasis was on transmission of knowledge from the teacher (who knew everything) to the learners (who were assumed to be empty vessels). The curriculum was objectives- d riven.

In South Africa, the Environmental Education Policy Initiative paved the way for the EE Curriculum Initiative (EECI) to develop a curriculum incorporating EE. In comparison, the Environmental and School Initiatives (ENSI) project is a n international project in EE co- ordinated by the Organisation for Economic Co- operation and Development's Centre for Educational Research and Innovation. Curriculum 2005 and outcomes - based education is comparable to Australian ENSI schools where EE refers to relevant issues rather than to knowledge. Investigation of such issues results in the emergence of knowledge and is therefore socially constructed.

Similar to Australia, South Africa bases its EE on the investigation of local environmental issues. Eight learning areas have been adopted by the South African Qualifications Authority (SAQA). In the new curriculum, EE has been identified as a phase organizer to be integrated into the eight learning areas where possible. But it is generally integrated into Human and Social Sciences (HSS) and Natural Sciences (NS), since it is "... the study of relationships between people, and between people and their environment". In the state of Queensland of Australia, four of the eight curriculum areas have been identified as essential to be incorporated into EE. In South Africa Environmental Studies is recommended at the Foundation Phase, Life Science/Education for Sustainable Living in the Senior Phase and Environmental Studies for FET. According to policy option 3 of the EECI, EE is best suited to FET and Adult Basic Education and Training (ABET) as a separate subject.

Australian Approach vs. South African Approach

In Australia, a single subject approach is not common within the compulsory ten years of schooling. Environmental studies merely encourages school participation in environmental projects where practical experience is vital for acquiring knowledge. In South Africa, the EECI states that EE could form a component of learning areas. In Australia it could also be a theme in any subject area. Another approach adopted by Australia is the faculty approach, where a faculty or year level adopts an environmental emphasis throughout the teaching of its subjects. In theory the South African and Australian policies are similar as EE forms part of pre-service teacher training. In-service education also includes action research in the development of resources. An action- based method is common to both approaches. Timetable organization inhibits EE implement ation in South Africa. Victorian policy, on the other hand recommends flexibility.

Both policies emphasize process evaluation. Thus, assessment is continuous, skill- based, conceptualized and incorporates EE principles. The emphasis is on formative and qualitative evaluation and focuses specifically on performance. Evaluation is holistic, interdisciplinary and builds an awareness of local and global significance. Apart from this, the Australian policy also looks for changes in students' attitudes and behaviour, as well as directions of any change. This is visible in the responsible actions learners take within their immediate home and school environment. The EEPI recommends that all institutions adopt an environmental policy which is common practice in Victoria and Queensland in Australia. The EEPI has developed a School Environmental Policy and Management Plan resource pack for South African schools. This policy is a statement of intentions and principles for improving a school's environmental performance. The policy development process encourages schools to audit existing activities and to formulate, evaluate and review EE goals and actions for key curriculum and extra -mural activities. In both countries all initiatives and programmes in EE focus on education for a sustainable future.This perception of the future suggests what knowledge, skills and values are central to EE.

Knowledge, Skills and Values in Environmental Education

Like the society it serves and will shape, today's education is in tra nsition, searching to identify elements of change, to preserve the cornerstones of traditional values. Many documents on EE, like Reorienting Teacher Education towards Sustainability: An Action Research Network Approach, refer to sustainability as the ability to reflect, to remember, to set contexts and to develop viable plans so that the transition can be made as smoothly as possible into a more sustainable 21st century. This perception of the future suggests that relevant knowledge; skil ls and values are central to education for sustainable development.

A perspective for reflection on the implications of education for a sustainable future and for curriculum changes to attain it, is provided in the framework of knowledge, skills and values as suggested by the Learning for a Sustainable Future. This is a non- profit Canadian organization whose mandate responds to the recommendations of the United Nations Conference on the Environment and Sustainable Development, held in Rio de Janeiro in 1992. The framework incorporates the idea that change will continue to take place all over the world and that we must constantly reassess the knowledge, skills and values that should be learnt in EE.

Knowledge

Knowledge in the following areas is essential for sustainable education:

— Planet earth as a finite system and the elements that constitute the planetary environment.

— The resources of the earth, particularly soil, water, minerals etc., their distribution and their role in supporting living organisms.

— The nature of ecosystems and biomes, their health and their interdependence within the biosphere.

- The dependenc e of humans on the environmental resources for life and sustenance.
- The sustainable relationship of native societies to the environment.
- The implications of resource distribution in determining the nature of societies and the rate and character of economic development.
- Characteristics of the development of human societies including nomadic, hunter - gatherer, agricultural, industrial and postindustrial, and the impact of each on the natural environment.
- The role of science and technology in the development of societies and the impact of these technologies on the environment.
- Philosophies and patterns of economic activity and their different impacts on the environment, societies and cultures.
- The process of urbanization and the implications of de - ruralisation.
- The interconnectedness of present world political, economic, environmental and social issues.
- Aspects of different perspectives and philosophies concerning the ecological and human environments.
- Co- operative international and national efforts to find solutions to common global issues and to implement strategies for a more sustainable future.
- The implications for the global community of the political, economic and socio-cultural changes needed for more sustainable future.
- Processes of planning, policy-making for action towards sustainability by governments, non- governmental organizations and the general public.

Skills

The following skills are seen as key to EE:

- Framing appropriate questions to guide relevant study and research.
- Defining such fundamental concepts as environment, community development and technology, and applying definitions to local, national and global experience.
- Using a range of bias and evaluating different points of view.
- Assessing the nature of bias and evaluating different points of view.

- Developing hypotheses based on balanced information, critical analysis and careful synthesis, and testing them against new information and personal experience and beliefs.
- Communicating information and viewpoints effectively.
- Working towards negotiated consensus and co- operative resolution of conflict.
- Developing co- operative strategies for appropriate action to change present relationships between ecological preservation and economic development.

Values

The following values are vital to EE:

- An appreciation of:
  - the resilience, fragility and beauty of nature and the interdependence and equal importance of all life forms;
  - the dependence of human life on the resources of a finite planet;
  - the role of human ingenuity and individual creativity in ensuring survival and the search appropriate and sustainable progress;
  - power of human beings to modify the environment;
  - the challenges faced by the human community in defining the processes needed for sustainability and in implementing the changes needed; and
  - the importance and the worth of individual responsibility and action.
- A sense of self-worth and rooted- ness in one's own culture and community.
- A respect for other cultures and recognition of the interdependence of the human community.
- A global perspective and loyalty to the world community.
- A concern for disparities and injustices, a commitment to human rights, and to the peaceful resolution of conflict.
- A sense of balance in deciding among conflicting priorities.
- Personal acceptance of a sustainable lifestyle and a commitment to participation in change.

— A realistic appreciation of the urgency of challenges facing the global community and the complexities that demand long-term planning for building a sustainable future.

International Initiatives

After more than 30 years of formal international efforts to promote environmental education, mainly through the UNESCO-UNEP International Environmental Education Program, concrete initiatives towards the institutional-isation of environmental education in the school system can now be observed. One of the main features of the actual international educational reform movement is the integration of environment in the curricula. This integration takes diverse forms. In some cases, environment is considered as a transversal theme or field of competencies, which cuts across the academic disciplines but which also calls for a specific educational transdisciplinary "niche".

In other cases, environmental education is explicitly mentioned as a specific and fundamental dimension of a global educational project. But often, according to the more recent UNESCO proposal of Education for a sustainable future, environment is presented as a theme amongst others in the integrating framework of education for sustainable development. It appears that in the actual international context characterised by a security crisis and by the globalisation process associated with the economization of human activities, the recent recognition of the legitimacy of environmental education (named as such or evoked) as an essential part of a holistic educational process, is facilitated when associated with the idea of sustainable development.

However it is looked at, the introduction of an environmental dimension in the curricula is a demanding task for the countries. The inspiration and beacons of international proposals to this effect has always been and is more than ever important, especially if there is a need or a desire to find strategic or financial support from leading international organisations in the field or to legitimate the initiatives and get help from national organisations.

Defining education is not an easy task and it could tend to fix and reify some ideas that need to remain dynamic and open. However not defining education can also lead to undesired results. It can be observed that most of the documents analysed do not propose a definition of education. Instead, they identify an end, an urgent end in fact, and then affirm that education must be reformed, urgently reformed, to serve such an end. Basically, education is thus an instrument to a predetermined finality.

In the proposals analysed, education is essentially instrumental into the service of environmental and developmental ends and almost no invitation is made to discuss those ends. Education is presented as an instrument to solve problems and to act directly upon these problems. All of the educational systems around the world are invited and expected to be reformed for such a purpose.

When the documents do define education, they usually provide a short humanistic definition which runs counter to pages and pages of prescriptions about how to educate in order to act on an environment mostly reduced to problems of resources that need to be better managed. "Agenda 21" thus asserts that "education, including formal education, public awareness and training should be recognised as a process by which human beings and societies can reach their fullest potential". The Declaration of Thessaloniki also states a view of education somewhat consistent with the humanistic dream:

> Education is an indispensable means to give all women and men in the world the capacity to own their own lives, to exercise personal choice and responsibility, to learn throughout life without frontiers, be they geographical, political, cultural, religious, linguistic or gender.

However it is difficult to reconcile such humanistic definitions with Agenda 21's more than two hundred and thirty action principles which form the heart of the proposal. In fact, the chapter on education, chapter 36, is embedded in the section whose revealing title, "Means of implementation", also contains chapters such as "Financial resources and mechanisms", "Transfer of environmentally sound technology ...", "Science for sustainable development", etc. These instrumental trends are most evident in recent proposals that deal with the training of people reduced - or elevated, depending on the vision adopted - to the status of "human resources" and "human capital". This aspect was however manifest right from the beginning where the Belgrade Charter asserts that "the reform of educational processes and systems is central to the building of this new development ethic and world economic order". In the same vein, the "International Strategy for Action in the Field of Environmental Education and Training for the 1990's" asserts at the outset that:

> In recent years there has been a gradual awareness, both world wide and within each individual State, of the role to be played by education in understanding, preventing and solving environmental problems. (UNESCO-UNEP, 1987)

The environment, conceived as a set of problems and as a reservoir of resources, is to be addressed within the boundaries of a developmental framework, thus as a support for development. Should not international proposals open up to some other possibilities of representation of the environment such as "environment as nature", "environment as a place to live and dwell" or "environment as a shared community project to be conducted in a critical and a political perspective".

Specifically within an educational context and perspective, should not one of the first goals of environmental education be to open up to explore the various ideas about what is that contemporary notion of environment. Where does this notion "environment" come from? What are its different representations and meanings? Who says what and who acts how? Why?

Numerous dimensions of relating to the environment are thus shadowed by this sort of economic approach to our relations to the surrounding world as highlighted in the proposals analysed. Just to name a few of the possible alternative approaches, one can think of place based approaches, literary approaches, artistic approaches, spiritual approaches, psychological approaches and the possible combination of these such as the place based and literary approaches to environmental education that are conveyed by organisations such as the "Orion Society" and the "Association for the Study of Literature and Environment".

Some of these approaches stress the importance of not focusing only on working on an environment "out-there", but also to work on the relationship and the inner dimensions of the relationships. The proposals reviewed, while often acknowledging in a sentence some of these dimensions and some of the alternative approaches, tend to downplay them and focus on the need to manage environmental problems of resources. There is none or very little consideration for the epistemological, philosophical, spiritual, psychological or physiological aspects of our relations to the world.

Another characteristic of the proposals is their manifest anthropocentrism as bear witness the title of the section of "Agenda 21" that addresses environmental issues, "Conservation and management of resources for development". The fourteen chapters of that section are subsumed under such a heading. The biosphere is destined to serve us, is destined for development through improving the productivity of its resources. Other ways of relating to the land, such as Aldo Leopold's land ethic of 1949, do not seem to be part of the Agenda. Other tensions and issues related to the differences within and between classes, age groups, cultures and nations are also subsumed in a global occidental anthropocentrism encompassing the whole humanity. Furthermore, the proposals, while rapidly acknowledging the importance of social issues, rely heavily on sciences and more specifically on environmental sciences and technological transfers as keys to the solution of environmental problems and thus to environmental education.

Even when the proposals stress the importance of taking into account society, environment, economy and development, and their integration, the proposal do not dwell at length on this and the ways in which it can be done. And social and psychological sciences are mostly seen as means of mobilizing people. A recurrent pattern can be observed in the successive proposals, which take the same historical view to assess the situation. The reports take into account the progress made since the last conference or report. The pattern is fixed in the following manner:

1) There have been good efforts made.
2) However the results are not sufficient.
3) The situation is thus degrading.
4) There is an emergency.

From there on, it is easy to understand that the proposals tend to fidget with impatience for environmental changes. This could explain why the proposals tend to focus more and more on actions, results, competencies and behavior changes while neglecting reflexivity and critical thinking. The Declaration of Thessaloniki thus stresses that:

> in order to achieve sustainability, an enormous co-ordination and integration of efforts is required in a number of crucial sectors and rapid and radical change of behaviours and lifestyles, including changing consumption and production patterns.

Another frequent pattern manifest in the proposals reviewed is their stressing on the crucial necessity of enrolling children for the environment. However, the reverse idea of providing diverse and healthy environments and providing specific environmental experiences for children development is generally absent. The notion of developmentally appropriate curriculum (referring to human personal and social development) is also noticeably absent.

Finally, as for the searches for a key or a root problem and for a total solution to the problem, the proposals tend to identify poverty as the main problem and to look at development and growth as the main solution. Here again a thread runs from the Belgrade Charter to the Thessaloniki Conference.

This thread is rather well subsumed in the 1987 Environmental Education and Training Congress, asserting that "in most developing countries, regardless of the region to which they belong, the basic problem is one of dire poverty, which in turn leads to deterioration of natural resources". Rarely are the causes of poverty researched nor is explored the notion that development and growth could be leading factors of social and environmental problems.

Every proposal calls for some sort of economic growth to solve environmental problems. Again, this theme runs from the Belgrade Charter calling for "measures that will support the kind of economic growth which will not have harmful repercussions on people; that will not in any way diminish their environment and their living conditions", to the "Rio Declaration" principle asserting that "states should cooperate to promote a supportive and open international economic system that would lead to economic growth and sustainable development in all countries, to better address the problems of environmental degradation".

Sustainable Development and Economic Growth

The integration of environmental concerns into development issues was strongly put forward by the 1987 report from the World Commission on Environment and Development. This preoccupation was already manifest in the Belgrade conference of 1975 and in the Tbilisi conference of 1977. Each of the conference called for a new form of development, but always somewhat embedded in some sort of economic growth.

In 1974, the United Nations "Declaration on the Establishment of a New International Economic Order" came out of "a special session of the General Assembly to study for the first time the problems of the raw materials and development, devoted to the consideration of the most important economic problems facing the world community". Such a statement indicates a view of the biosphere as a reservoir to serve the development of economic growth. It illustrates how the environment, "problems of the raw materials", is linked to development and how they are both subsumed as "most important economic problems".

The International Environmental Education Program (IEEP) launched in 1975 is firmly rooted in this establishment of a new economic order where the environment must be preserved for the raw materials so as to afford a continuous and equitable development throughout the world. As for the concept of education, the notion of development is generally ill defined in the documents analysed. Rarely is there a formal definition. However, since the issues of the environment are being framed in an international perspective, poverty is seen as the major problem to be addressed and development is looked at as a key element of solution.

The lumping of environment and development in this perspective of an emerging global market thus calls for economic growth to eliminate poverty. This integration of the environment within development, as a necessary condition and an "incontournable" constraint, overshadows representations of the environment as something other than a problem of resources for economic development. Stepping back to reflect on educational practices taking into account the surrounding world, it is manifest that environment and development can sometimes be dissociated, even more so in an educational perspective. This fusing of the environment with the idea of development now defined as not dissociable, can very easily hide the specific contexts of environmental abuse and of developmental problems. It should be possible to explore in depth environmental issues from a variety of perspectives, sometimes with a developmental perspective, sometimes without.

The notion of development suffers somewhat the same fate as the notions of education and environment. In the same way that the humanistic vision of education or total vision of the environment forwarded by the proposals tend to collapse respectively under the weigh of the instrumental view of education and the view of the environment as resources, the notion of development tends to bear the weigh of economic growth. The Belgrade Charter loudly calls for new patterns of development.

> The recent United Nations Declaration for a New International Economic Order calls for a new concept of development - one which takes into account

> the satisfaction of the needs and wants of every citizen on earth, of the pluralism of societies and of the balance and harmony between humanity and the environment. What is called for is the eradication of the basic causes of poverty, hunger, illiteracy, pollution, exploitation and domination. The previous pattern of dealing with these crucial problems on a fragmentary basis is no longer workable.

The last sentence of this call has a familiar ring since it is the call for the idea of integration, explored in the following section, that was to climax with the Brundtland report and "Agenda 21". However strong the call "for a new concept of development" in the Belgrade Charter, the next paragraph of the same 1976 document asserts that "the resources of the world should be developed in ways which will benefit all of humanity and provide the potential for raising the quality of life for everyone. Such a goal bears the weigh of an economic view of development.

There is then a tension in the notions of development forwarded by the proposals reviewed: on the one hand a vision of human development and on the other hand a view of resources development, of economic development and of economic growth. Even the single terms of the tension are ill defined and thus open the way to very diverse and sometimes opposite interpretation. What is meant by quality of life? What is meant by human development? In such a confusion, gauging development with the indicators of economic growth seems to exert a strong appeal.

The tension between the two discourses concerning development caries through the proposals and is manifest again in the Declaration of Thessaloniki.

> Poverty makes the delivery of education and other social services more difficult and leads to population growth and environmental degradation. Poverty reduction is thus an essential goal and indispensable condition of sustainability.

Economic growth can thus easily appear as a major solution. The reasoning can all too easily become the following: there is a need for economic development to eliminate poverty and sustain human development. From then on, there is an easy perversion to the idea that human development is based on economic growth. And since environmental and developmental issues are not dissociable, a new form of economic growth will also solve environmental problems of resources. But this "new form" of economic growth or development is very poorly characterised. References to endogenous (or local, or alternative) development are shadowed by the preoccupation of managing environment as a backstore of "raw material". Even more overshadowed is the idea, forwarded by Sachs for example, that development in and by itself is a problem and that such an idea, should be discarded in the perspective of a postdevelopment era.

Subsuming environmental and developmental issues with sustained economic growth appears more bluntly within the "United Nations Programme of Action of the International Conference on Population and Development". In the document, calls for "sustained economic growth in the context of sustainable development" or for "sustained economic growth and sustainable development" appear more than twenty times. The first item in the preamble subsumes, in only five sentences, so many themes highlighted in the discourse.

The 1994 International Conference on Population and Development occurs at a defining moment in the history of international cooperation. With the growing recognition of global population, development and environmental interdependence, the opportunity to adopt suitable macro- and socio-economic policies to promote sustained economic growth in the context of sustainable development in all countries and to mobilize human and financial resources for global problem-solving has never been greater. Never before has the world community had so many resources, so much knowledge and such powerful technologies at its disposal which, if suitably redirected, could foster sustained economic growth and sustainable development. None the less, the effective use of resources, knowledge and technologies is conditioned by political and economic obstacles at the national and international levels. Therefore, although ample resources have been available for some time, their use for socially equitable and environmentally sound development has been seriously limited.

Development is thus presented as a right and an obligation strongly associated to economic growth. Such a proposal overshadows the idea of looking at development as an option, as a choice and minimally of exploring the notion of development and its different meanings. Is there not a fear that an endogenous or alternative vision of development could be a "political and economic obstacle" to the global sustainable growth solution? In an educational context, the exploration of the very contemporary and polysemic notion of development can be envisioned in a fashion similar to the exploration of the notion of environment. Where does this notion of development come from? What are its different meanings to different people?

All the proposals being framed in a paradoxical context of post cold war negotiation between delegates of nations, thus being "inter-national" and at the same time the new vision of the single finite earth seen from the moon "one-world", the authors insist on the need for cooperation, for solidarity and interdependency. In the shadow of this call for fraternity between delegates of nations striving for a common goal are the notions of autonomy, self-management, self-reliance. There would be an interest to question the inherent idea of solidarity in the sustainability proposal in light of the assertion found in the President's Council on Sustainable Development document following which education for sustainable development is a means to increase national competitiveness in a global economy.

Notion of Integration

One of the objectives of this research is to analyse how the international proposals address the idea of integration, which has been revealed by a simple statistic lexicometric treatment as a key word. For now, at least six aspects or meanings have been uncovered for the notion of integration. They are more manifest with "Agenda 21" which seem to be a climax call to integrate. A first facet of integration has to do with the educational system. In the proposals, environmental education has to be integrated in all disciplines, in every program, at all levels. This is the multidisciplinary and multisector view of integration. This is manifest in 1987, in the Moscow Congress, "it has been clear that environmental education should be a dimension of all subjects and areas of education", and again in 1992, in Chapter 36 of "Agenda 21", "to be effective, environment and development education (...) should be integrated in all disciplines".

There would also be a need to insist on the necessity to introduce environmental policies and practices in the planning and management of physical infrastructures and with daily life activities of educational institutions: schools, colleges, universities, etc. While this preoccupation is sometimes apparent in some of the proposals, it is not strongly highlighted but rather overshadowed. The academic discourse has to be coherent with the "hidden curriculum" of the institution, taking into account the "pedagogy of place" as highlighted by Orr.

A second aspect of integration becomes rapidly evident in that all the proposals stress the importance of interdisciplinary approaches. While acknowledging the difficulty of it, it is wholeheartedly recommended. Such an interdisciplinary view of integration runs from the Belgrade Charter "Guiding Principles of Environmental Education Programmes" stating "environmental education should be interdisciplinary in its approach" to the Thessaloniki Declaration stating "addressing sustainability requires a holistic, interdisciplinary approach which brings together the different disciplines and institutions while retaining their different identities". This is the second aspect of integration: the integration of academic disciplines for EE. It may be observed here that knowledge is necessarily considered in reference to the modern notion of discipline, may it be inter-discipline. There are few mentions of the possible integration of diverse types of knowledge that are not yet or rarely legitimated, such as traditional knowledge or experiential knowledge. When traditional knowledge is valued, it is generally for its potential to sustain and promote the worldview forwarded by the proposal, not for its own intrinsic value.

A third aspect of integration has to do with the extension of environmental education outside of the formal educational system. Environmental education has to be extended and integrated to every social milieu, in every profession, throughout the world and in a life long process. Such a notion of integration starts with the Belgrade Charter identification of audiences and is manifest in every proposal such as in the program of action plan stemming from the Conference on population and development and in the Declaration of Thessaloniki:

> Greater public knowledge, understanding and commitment at all levels, from the individual to the international, are vital to the achievement of the goals and objectives of the present Programme of Action. In all countries and among all groups, therefore, information, education and communication activities concerning population and sustainable development issues must be strengthened. The reorientation of education as a whole towards sustainability involves all levels of formal, non formal and informal education in all countries.

In this level of integration, there is a need to insist on the idea of achieving a real educational society, operationalized in educational communities, where the different actors cooperate to play a specific and complementary role in the shared essential responsibility of engaging in environmental education, adopting a critical approach to socio-environmental realities. Such a perspective is clearly different from mobilizing populations for predetermined goals, which seems to lie at the core of many proposals. As for mobilizing the media for environmental issues, one can doubt the ease with which such a goal may be achieved in some context where the media are strongly dependent on revenues coming from a sort of runaway growth economy.

A fourth view of integration deals with the notion of the environment. As noted earlier, the proposals assert that environmental education has to consider the totality of the environment however that totality be defined. Is it not problematic to consider environment as such a big whole that there is no more specificity or angle to grasp it?

A fifth meaning to the notion of integration appears more drastically with the Rio Declaration. Environment and development are now said to have to be integrated together with some other issues such as peace, health and others. Within the body of the proposals reviewed, it is with the Rio conference that environment and development become glued together within typical formulas.

The following excerpts from the beginning of Chapter 36 of "Agenda 21" are typical. The emphasis on the key terms is added.

> Education is critical for promoting sustainable development and improving the capacity of the people to address environment and development issues. While basic education provides the underpinning for any environmental and development education, the latter needs to be incorporated as an essential part of learning. Both formal and non-formal education are indispensable to changing people's attitudes so that they have the capacity to assess and address their sustainable development concerns. It is also critical for achieving environmental and ethical awareness, values and attitudes, skills and behaviour consistent with sustainable development and for effective public participation in decision-making. To be effective, environment and development education should deal with the dynamics of both the physical/biological and socio-economic environment and human (which may include spiritual) development, should be integrated in all disciplines, and should employ formal and non-formal methods and effective means of communication.

The Rio Declaration on Environment and Development states, in "Principle 25" that "Peace, development and environmental protection are interdependent and indivisible". "Agenda 21" and the Rio Declaration contain no less than 150 mentions of "environment and development". In a little more than 3500 words that Chapter 36 contains, "environment" is linked to "development" 35 times. Once in every 100 words.

The Declaration of Thessaloniki also builds on this multidimensional view of integration when it states that "sustainability encompasses not only environment but also poverty, population, health, food, security, democracy, human rights and peace".

The importance of considering the integration of these different aspects in a holistic and systemic understanding of the realities, should not bring though to embrace all of a sudden and unreflexivly a disproportionate educational task, where the mixing up of a huge quantity of diverse and specific objectives could lead to confusion and inefficacy. It would be important to avoid the sinking of the disquieting and demanding environmental preoccupation in a single overbooked agenda.

Finally, a sixth aspect of integration has a clear economic bent. The solution to the environmental and developmental problems calls for the integration of all countries in a world economy and international trade system. Here is the integration within globalization.

> Economic integration processes have intensified in recent years and should impart dynamism to global trade and enhance the trade and development possibilities for developing countries. In recent years, a growing number of these countries have adopted courageous policy reforms involving ambitious autonomous trade liberalization, while far-reaching reforms and profound restructuring processes are taking place in Central and Eastern European countries, paving the way for their integration into the world economy and the international trading system.

Integration really sounds as a key word in the proposals. Climaxing with the Rio and Cairo reports, such a call for a total integration justifies the title of this section: integration as a macrostrategy. Of course, this idea of integration may be fruitful, even necessary in some respects. It is a way to avoid the breakdown of complex realities and to search for a global vision of these realities, so as to take pertinent decisions. But in the shadow of such a multidimensional and global call to integrate almost everything is the nitty-gritty task of concretely trying to do some integration to address, let us not forget, the environmental issues originally lying at the base of this whole international scaffolding.

What does one integrate? why? How does one integrate? What are the limits to integration? Is not a total and global call for integration an omnipotent dream? Is there not the danger that such an integration necessarily calls for a lowest common denominator? Could such a desire of global integration call for a lowest common denominator that will paradoxically cause environmental and social disintegration? In this perspective, it seems that the expression "Think globally, act locally" has been the crucible of an important confusion. It consists in attaining a complete, systemic and global vision of a truly complex object (environment, development), in order to act in an appropriate way. Now, this object, this reality can very well be situated at the local level, the level of the community.

The invitation to global thinking (holistic) does not necessarily mean the adoption of a planetary or international perspective, or of an "integrating" global whole. As noted by Esteva and Prakash, such a perspective is neither always attainable nor always appropriate. Global thinking (holistic) applies first locally, at this first scale where becomes possible a real knowledge and the mobilization of the power to act linked to the exercise of an intrinsic responsibility.

There is a common thread linking the global conferences of Stockholm, Belgrade, Tbilisi, Moscow, Rio de Janeiro, Cairo and Thessaloniki. One element of this thread is to call in education to implement a program of action to reform the world towards sustainability. This can very easily overshadow diverse educational approaches to experiencing the world and reflecting upon the environment and our own relation to it. The proposals problematically forward a view of the environment as problems of resources and a view of development as mainly associated with economic growth presented as a condition to human development.

References

Hungerford, H. R. et.al. (eds.), (2001). *Essential Readings in Environmental Education* (2nd ed., pp. 17-31). Champaign, IL: Stipes Publishing.

Engleson, D. C., & Yockers, D. H., (1994). *EA guide to curriculum planning in environmental education.* Madison, WI: Wisconsin Department of Public Instruction.

Swan, James A. and William B. Stapp, (1974). *Environmental Education: Strategies Toward a More Livable Future.* New York: Sage Publications.

Vivian, V. Eugene, (1973). *Sourcebook for Environmental Education*, Saint Louis: C.V. Mosby Company.

Bibliography

Begon, M.; Townsend, C. R.; Harper, J. L. (2005). *Ecology: From Individuals to Ecosystems* (4th ed.). Wiley-Blackwell. p. 752. ISBN 1-4051-1117-8.

Brasseur, Guy P.; Orlando, John J.; Tyndall, Geoffrey S. (1999). *Atmospheric Chemistry and Global Change*. Oxford University Press.

Brasseur, Guy P.; Orlando, John J.; Tyndall, Geoffrey S. (1999). *Atmospheric Chemistry and Global Change.* Oxford University Pres

Carroll, B. and Turpin T. (2009). *Environmental impact assessment handbook, second edition* Thomas Telford Ltd.

Conacher, Arthur; Conacher, Jeanette (1995). *Rural Land Degradation in Australia*. South Melbourne, Victoria: Oxford University Press Australia. p. 2.

Davic, R. D. (2003). "Linking keystone species and functional groups: a new operational definition of the keystone species concept". *Conservation Ecology* 7 (1): r11.

Davis, Devra (2002). *When Smoke Ran Like Water: Tales of Environmental Deception and the Battle Against Pollution.* Basic Books

Engleson, D. C., & Yockers, D. H., (1994). *EA guide to curriculum planning in environmental education.* Madison, WI: Wisconsin Department of Public Instruction.

Eswaran, H.; R. Lal and P.F. Reich. (2001). "Land degradation: an overview". *Responses to Land Degradation. Proc. 2nd. International Conference on Land Degradation and Desertification.* New Delhi, India: Oxford Press. Retrieved 2006-06-20.

Finlayson-Pitts, Barbara J.; Pitts, James N., Jr. (2000). *Chemistry of the Upper and Lower Atmosphere*. Academic Press.

Glasson, J; Therivel, R; Chadwick A, (2005). *Introduction to Environmental Impact Assessment*, Routledge, London.

Grumbine, R. E. (1994). "What is ecosystem management?". *Conservation Biology* 8 (1): 27–38.

Hammond, H. (2009). *Maintaining Whole Systems on the Earth's Crown: Ecosystem-based Conservation Planning for the Boreal Forest*. Slocan Park, BC: Silva Forest Foundation. p. 380.

Hanna, K; (2009). *Environmental Impact Assessment: Practice and Participation* Second edition, Oxford,

Holling, C. S. (2004). "Understanding the complexity of economic, ecological, and social systems". *Ecosystems* 4 (5): 390–405.

Hungerford, H. R. et.al. (eds.), (2001). *Essential Readings in Environmental Education* (2nd ed., pp. 17-31). Champaign, IL: Stipes Publishing.

Ian Sample (2007-08-31). "Global food crisis looms as climate change and population growth strip fertile land". *The Guardian*. Retrieved 2008-07-23.

Johnson, Douglas; Lewis, Lawrence. (2007), *Land Degradation; Creation and Destruction*, Maryland, USA

Kormondy, E. E. (1995). *Concepts of Ecology* (4th ed.). Benjamin Cummings.

McCann, K. (2007). "Protecting biostructure". *Nature* 446 (7131): 29. Bibcode 2007Natur.446...29M.

Mclntosh, R. P. (1985). *The Background of Ecology: Concept and Theory*. Cambridge University Press. pp. 400.

Mills, L. S.; Soule, M. E.; Doak, D. F. (1993). "The keystone-species concept in ecology and conservation".*BioScience* 43 (4): 219–224.

Noss, R. F.; Carpenter, A. Y. (1994). *Saving Nature's Legacy: Protecting and Restoring Biodiversity*. Island Press. p. 443.

Odum, E. P.; Barrett, G. W. (2005). *Fundamentals of Ecology*. Brooks Cole. p. 598.

Petts, J. (ed), *Handbook of Environmental Impact Assessment* Vol 1 & 2, Blackwell, Oxford.

Risse-Kappen, T (1995). *Bringing transnational relations back in: non-state actors, domestic structures, and international institutions*. Cambridge: Cambridge University Press. pp. 3–34.

Seinfeld, John H.; Pandis, Spyros N. (2006). *Atmospheric Chemistry and Physics: From Air Pollution to Climate Change* (2nd Ed.). John Wiley and Sons, Inc.

Stockings, Mike; Murnaghan, Niamh. (2000), *Land Degradation - Guidelines for Field Assesment*, Norwich, UK, pp. 7–15

Swan, James A. and William B. Stapp, (1974). *Environmental Education: Strategies Toward a More Livable Future*. New York: Sage Publications.

Turner, D.B. (1994). *Workbook of atmospheric dispersion estimates: an introduction to dispersion modeling* (2nd ed.). CRC Press.

Vivian, V. Eugene, (1973). *Sourcebook for Environmental Education*, Saint Louis: C.V. Mosby Company.

Warneck, Peter (2000). *Chemistry of the Natural Atmosphere* (2nd Ed.). Academic Press.

Wayne, Richard P. (2000). *Chemistry of Atmospheres* (3rd Ed.). Oxford University Press.

Whittaker, R. H.; Levin, S. A.; Root, R. B. (1973). "Niche, habitat, and ecotope". *The American Naturalist* 107(955): 321–338.

Wilson, E. O. (1992). *The Diversity of Life*. Harvard University Press. pp. 440.